CLEP

College Level Examination Program

Algebra

Andy Gaus, MS
Kathleen Morrison, MS
Sujata S. Millick, PhD

XAMonline

Copyright © 2016
All rights reserved. No part of the material protected by this copyright notice may be reproduced or utilized in any form or by any means, electronic or mechanical, including photocopying or recording or by any information storage and retrievable system, without written permission from the copyright holder.

To obtain permission(s) to use the material from this work for any purpose including workshops or seminars, please submit a written request to:

XAMonline, Inc.
21 Orient Avenue
Melrose, MA 02176
Toll Free: 1-800-301-4647
Email: info@xamonline.com
Web: www.xamonline.com
Fax: 1-617-583-5552

Library of Congress Cataloging-in-Publication Data
Gaus, Andy

CLEP College Algebra / Andy Gaus
   ISBN: 978-1-60787-559-8

1. CLEP   2. Study Guides   3. Mathematics   4. Algebra

Disclaimer:
The opinions expressed in this publication are the sole works of XAMonline and were created independently from the College Board, or other testing affiliates. Between the time of publication and printing, specific test standards as well as testing formats and website information may change that are not included in part or in whole within this product. XAMonline develops sample test questions, and they reflect similar content as on real tests; however, they are not former tests. XAMonline assembles content that aligns with test standards but makes no claims nor guarantees candidates a passing score.

Cover photo provided by ©Can Stock Photo Inc./Tatiana53

Printed in the United States of America
CLEP College Algebra
ISBN: 978-1-60787-559-8

Table of Contents

About CLEP. 7
    I. The College Level Examination Program 7
    II. Approaching a College about CLEP. 10
    III. Preparing to Take CLEP Examinations 15
    IV. Interpreting Your Scores 18
About CLEP College Algebra 20
Order of operations 22
Real numbers 22
Equations and inequalities. 23
Inequalities 24
Properties of equations and inequalities 25
Matrices. 26
Addition, subtraction, and multiplication of matrices 26
Determinants of matrices. 27
Properties of operations 27
Summary of the properties of operations 28
Solving linear equations 28
Solving linear inequalities 29
Graphing a linear inequality. 31
Absolute value equations 33
Absolute value inequalities 34
Domain and range. 34
Functions. 35
Properties of functions. 36
Families of functions 37
Graphing a linear equation 37
Procedure for graphing a linear equation 37
Linear functions. 39
Direct variation. 39
Sequences and series. 40
Arithmetic series 40
Geometric series 42
Intercepts. 43

Systems of linear equations 44
Solving a system of equations............................. 44
Solving a system of equations by graphing.................. 45
Solving a system of equations by elimination 45
Solving a system of equations by substitution................ 46
Solving a system of linear inequalities...................... 48
Operations with exponents 49
Properties of logarithms................................... 50
Solving problems involving exponential or logarithmic functions... 52
Expanding polynomials 52
Factoring polynomials 53
- Removing common factors 53
- Factoring the difference of two squares 53

Quadratic equations and quadratic expressions 56
Solving quadratic equations 56
- Taking the square root................................ 56
- Solving quadratic equations by factoring................ 57
- Solving quadratic equations with the quadratic formula 57

Quadratic equations and imaginary numbers 58
Complex numbers 59
The rational root theorem 60
The complex conjugate root theorem 61
Quadratic inequalities 61
Modeling functions.. 62
Finding powers of a binomial (the binomial theorem) 65
Composition of functions.................................. 66
Inverses of functions 66
Finding inverses of functions............................... 66
Operations with radicals 67
Addition and subtraction of radicals......................... 68
Multiplication and division of radicals 68
Transformational geometry: translations, rotations,
reflections, and scaling 69
Symmetries 75
Piecewise functions....................................... 77

Inverse variation 77
Rational functions 78
Graphing rational functions using asymptotes. 79
Operations on rational expressions. 80
Simplifying rational expressions 81
Solving equations involving rational expressions 83
Polynomial functions 84
The fundamental theorem of algebra 85
The factor theorem 86
Factorials. 86
Permutations and combinations 86
Objects arranged in a row 88

Sample Test 1 89
Sample Test 1: Answers and Explanations 112

Sample Test 2 149
Sample Test 2: Answers and Explanations 172

Sample Test 3 211
Sample Test 3: Answers and Explanations 234

MEET THE AUTHORS:

Andy Gaus has created math content online and in print for such major publishers and content providers as Glencoe, Holt, Pearson, Aptara Corporation and Brown Educational Network. He is also a theater pianist who has played numerous shows in the Boston area, where he lives.

Kathleen Morrison is a certified Math and Science teacher licensed in the state of Illinois. Her classroom experience spans 16 years covering subject areas from Pre-Algebra to Pre-Calculus, Geometry to Trigonometry, along with interdisciplinary experience combining Algebra and Physical Science. She is currently working as a private mathematics tutor, a substitute teacher, and a writer of mathematics practice exercises.

Dr. Sujata Millick works at the intersection of the technology, education, security and privacy domains. She has over two decades of public sector executive and programmatic experience in science, R&D, and STEM education portfolios in the defense, commerce, and maritime organizations. She pursues work in the areas of education, internet connectedness, and emerging technologies.

I. The College-Level Examination Program

How the Program Works

CLEP exams are administered at over 1,800 institutions nationwide, and 2,900 colleges and universities award college credit to those who perform well on them. This rigorous program allows many self-directed students of a wide range of ages and backgrounds to demonstrate their mastery of introductory college-level material and pursue greater academic success. Students can earn credit for what they already know by getting qualifying scores on any of the 33 examinations.

The CLEP exams cover material that is taught in introductory-level courses at many colleges and universities. Faculty at individual colleges review the exams to ensure that they cover the important material currently taught in their courses.

Although CLEP is sponsored by the College Board, only colleges may grant credit toward a degree. To learn about a particular college's CLEP policy, contact the college directly. When you take a CLEP exam, you can request that a copy of your score report be sent to the college you are attending or planning to attend. After evaluating your score, the college will decide whether or not to award you credit for a certain course or courses, or to exempt you from them.

If the college decides to give you credit, it will record the number of credits on your permanent record, thereby indicating that you have completed work equivalent to a course in that subject. If the college decides to grant exemption without giving you credit for a course, you will be permitted to omit a course that would normally be required of you and to take a course of your choice instead.

The CLEP program has a long-standing policy that an exam may not be taken within the specified wait period. This waiting period provides you with an opportunity to spend additional time preparing for the exam or the option of taking a classroom course. If you violate the CLEP retest policy, the administration will be considered invalid, the score canceled, and any test fees will be forfeited. If you are a military service member, please note that DANTES will not fund retesting on a previously funded CLEP exam. However, you may personally fund a retest after the specified wait period.

The CLEP Examinations

CLEP exams cover material directly related to specific undergraduate courses taught during a student's first two years in college. The courses may be offered for three, four, six or eight semester hours in general areas such as mathematics, history, social sciences, English composition, natural sciences and humanities. Institutions will either grant credit for a specific course based on a satisfactory score on the related exam, or in the general area in which a satisfactory is earned. The credit is equal to the credit awarded to students who successfully complete the courses. See the Table of Contents for a complete list of all exam titles.

What the Examinations Are Like

CLEP exams are administered on computer and are approximately 90 minutes long, with the exception of College Composition, which is approximately 120 minutes long. Most questions are multiple-choice; other types of questions require you to fill in a numeric answer, to shade areas of an object, or to put items in the correct order. Questions using these kinds of skills are called zone, shade, grid, scale, fraction, numeric entry, histogram and order match questions.

CLEP College Composition includes a mandatory essay section, responses to which must be typed into the computer.

Some of the examinations have optional essays. You should check with the individual college or university where you are sending your score to see whether an optional essay is required for those exams. These essays are administered on paper and are scored by faculty at the institution that receives your score.

Where to Take the Examinations and How to Register

CLEP exams are administered throughout the year at over 1,800 test centers in the United States and select international sites. Once you have decided to take a CLEP examination, you can log into My Account at https://clepportal.collegeboard.org/myaccount to create and manage your own personal accounts, pay for CLEP exams and purchase study materials. You can self-register at any time by completing the online registration form.

Through My Account you can also access a list of institutions that administer CLEP and locate a test center in your area. **After paying for your exam through My Account, you must still contact the test center to schedule your CLEP exam.**

If you are unable to locate a test center near you, call 800-257-9558 for more information.

ACE's College Credit Recommendation Service

The College Credit Recommendation Service (CREDIT) of the American Council on Education (ACE) enables you to put all of your educational achievements on a secure and universally accepted ACE transcript. All of your ACE-evaluated courses and examinations, including CLEP, appear in an easy-to-read format that includes ACE credit recommendations, descriptions and suggested transfer areas. The service is perfect for candidates who have acquired college credit at multiple ACE-evaluated organizations or credit-by-examination programs. You may have your transcript released at any time to the college of your choice. There is a one-time setup fee of $40 (includes the cost of your first transcript) and a fee of $15 for each transcript requested after release of the first. ACE has an additional transcript service for organizations offering continuing education units.

The College Credit Recommendation Service is offered through ACE's Center for Lifelong Learning. For more than 50 years, ACE has been at the forefront of the evaluation of education and training attained outside the classroom. For more information about ACE CREDIT, contact:

ACE CREDIT
One Dupont Circle, NW
Suite 250
Washington, DC 20036

ACE's Call Center is open Monday to Friday, 8:45 a.m. to 4:45 p.m., and can be reached at 866-205-6267 or CREDIT@ace.nche.edu. Staff are able to assist you with courses and certifications that carry ACE recommendations for both civilian organizations and training obtained through the military.

If you are already registered for an ACE transcript, you can access your records and order transcripts using the ACE Online Transcript System: https://www.acenet.edu/transcripts/.

ACE's Center for Lifelong Learning can be found on the Internet at: http://www.acenet.edu/ higher-education.

How Your Score Is Reported

You have the option of seeing your CLEP score immediately after you complete the exam, except in the case of College Composition, for which scores are available four to six weeks after the exam date. Once you choose to see your score, it will be sent automatically to the institution you have designated as a score recipient; it cannot be canceled. You will receive a candidate copy of your score before you leave the test center. If you have tested at the institution that you have designated as a score recipient, it will have immediate access to your test results.

If you do not want your score reported, you may select that as an option at the end of the examination before the exam is scored. Once you have selected the option to not view your score, the score is canceled.

The score will not be reported to the institution you have designated, and you will not receive a candidate copy of your score report. You will have to wait the specified wait period before you can take the exam again.

CLEP scores are kept on file for 20 years. During this period, for a small fee, you may have your transcript sent to another college or to anyone else you specify. Your score(s) will never be sent to anyone without your approval.

II. Approaching a College about CLEP

The following sections provide a step-by-step guide to learning about the CLEP policy at a particular college or university. The person or office that can best assist you may have a different title at each institution, but the following guidelines will lead you to information about CLEP at any institution.

Adults and other nontraditional students returning to college often benefit from special assistance when they approach a college. Opportunities for adults to return to formal learning in the classroom are now widespread, and colleges and universities have worked hard to make this a smooth process for older students. Many colleges have established special offices that are staffed with trained professionals who understand the kinds of problems facing adults returning to college. If you think you might benefit from such assistance, be sure to find out whether these services are available at your college.

How to Apply for College Credit

STEP 1. *Obtain, or access online, the general information catalog and a copy of the CLEP policy from each college you are considering.*

Information about admission and CLEP policies can be obtained on the college's website at clep.collegeboard.org/search/colleges, or by contacting or visiting the admissions office. Ask for a copy of the publication in which the college's complete CLEP policy is explained. Also, get the name and the telephone number of the person to contact in case you have further questions about CLEP.

STEP 2. *If you have not already been admitted to a college that you are considering, look at its admission requirements for undergraduate students to see whether you qualify.*

Whether you're applying for college admission as a high school student, transfer student or as an adult resuming a college career or going to college for the first time, you should be familiar with the requirements for admission at the schools you are considering. If you are a nontraditional student, be sure to check whether the school has separate admissions requirements that might apply to you. Some schools are very selective, while others are "open admission."

It might be helpful for you to contact the admissions office for an interview with a counselor. State why you want the interview and ask what documents you should bring with you or send in advance. (These materials may include a high school transcript, transcript of previous college work or completed application for admission.) Make an extra effort to have all the information requested in time for the interview.

During the interview, relax and be yourself. Be prepared to state honestly why you think you are ready and able to do college work. If you have already taken CLEP exams and scored high enough to earn credit, you have shown that you are able to do college work. Mention this achievement to the admissions counselor because it may increase your chances of being accepted. If you have not taken a CLEP exam, you can still improve your chances of being accepted by describing how your job training or independent study has helped prepare you for college-level work. Discuss with the counselor what you have learned from your work and personal experiences.

STEP 3. *Evaluate the college's CLEP policy.*

Typically, a college lists all its academic policies, including CLEP policies, in its general catalog or on its website. You will probably find the CLEP policy statement under a heading such as Credit-by-Examination, Advanced Standing, Advanced Placement or External Degree Program. These sections can usually be found in the front of the catalog. You can also check out the institution's CLEP Policy by visiting clep.collegeboard.org/search/colleges.

Many colleges publish their credit-by-examination policies in separate brochures, which are distributed through the campus testing office, counseling center, admissions office or registrar's office. If you find a very general policy statement in the college catalog, seek clarification from one of these offices.

Review the material in the section of this chapter entitled "Questions to Ask about a College's CLEP Policy." Use these guidelines to evaluate the college's CLEP policy. If you have not yet taken a CLEP exam, this evaluation will help you decide which exams to take. Because individual colleges have different CLEP policies, a review of several policies may help you decide which college to attend.

STEP 4. *If you have not yet applied for admission, do so as early as possible.*

Most colleges expect you to apply for admission several months before you enroll, and it is essential that you meet the published application deadlines. It takes time to process your application for admission. If you have yet to take a CLEP exam, you may want to take one or more CLEP exams while you are waiting for your application to be processed. Be sure to check the college's CLEP policy beforehand so that you are taking exams your college will accept for credit. You should also find out from the college when to submit your CLEP score(s).

Complete all forms and include all documents requested with your application(s) for admission.

Normally, an admission decision cannot be reached until all documents have been submitted and evaluated. Unless told to do so, do not send your CLEP score(s) until you have been officially admitted.

STEP 5. *Arrange to take CLEP exam(s) or to submit your CLEP score(s).*

CLEP exams can be taken at any of the 1,800 test centers world-wide. To locate a test center near you. clep.collegeboard.org/search/test-centers.

If you have already taken a CLEP exam, but did not have your score sent to your college, you can have an official transcript sent at any time for a small fee. Fill out the Transcript Request Form included on the same page as your exam score. If you do not have the form, visit clep.collegeboard.org/about/score to download a copy, or call 800-257-9558 to order a transcript using a major credit card. Completed forms should be faxed to 610-628-3726 or sent to the following address, along with a check or money order made payable to CLEP for $20 (this fee is subject to change).

CLEP Transcript Service
P.O. Box 6600
Princeton, NJ 08541-6600

Transcripts will only include CLEP scores for the past 20 years; scores more than 20 years old are not kept on file.

Your CLEP scores will be evaluated, probably by someone in the admissions office, and sent to the registrar's office to be posted on your permanent record once you are enrolled. Procedures vary from college to college, but the process usually begins in the admissions office.

STEP 6. *Ask to receive a written notice of the credit you receive for your CLEP score(s).*

A written notice may save you problems later, when you submit your degree plan or file for graduation. In the event that there is a question about whether or not you earned CLEP credit, you will have an official record of what credit was awarded. You may also need this verification of course credit if you meet with an academic adviser before the credit is posted on your permanent record.

STEP 7. *Before you register for courses, seek academic advising.*

A discussion with your academic adviser can help you to avoid taking unnecessary courses and can tell you specifically what your CLEP credit will mean to you. This step may be accomplished at the time you enroll. Most colleges have orientation sessions for new students prior to each enrollment period. During orientation, students are usually assigned academic advisers who then give them individual help in developing long-range plans and course schedules for the next semester. In conjunction with this counseling, you may be asked to take some additional tests so that you can be placed at the proper course level.

Questions to Ask about a College's CLEP Policy

Before taking CLEP exams for the purpose of earning college credit, try to find the answers to these questions:

1. *Which CLEP exams are accepted by the college?*

   A college may accept some CLEP exams for credit and not others — possibly not the exams you are considering. For this reason, it is important that you know the specific CLEP exams for which you can receive credit.

2. *Does the college require the optional free-response (essay) section for exams in composition and literature as well as the multiple-choice portion of the CLEP exam you are considering? Will you be required to pass a departmental test such as an essay, laboratory or oral exam in addition to the CLEP multiple-choice exam?*

   Knowing the answers to these questions ahead of time will permit you to schedule the optional free-response or departmental exam when you register to take your CLEP exam.

3. *Is CLEP credit granted for specific courses at the college? If so, which ones?*

   You are likely to find that credit is granted for specific courses and that the course titles are designated in the college's CLEP policy. It is not necessary, however, that credit be granted for a specific course for you to benefit from your CLEP credit. For instance, at many liberal arts colleges, all students must take certain types of courses; these courses may be labeled the core curriculum, general education requirements, distribution requirements or liberal arts requirements. The requirements are often expressed in terms of credit hours. For example, all students may be required to take at least six hours of humanities, six hours of English, three hours of mathematics, six hours of natural science and six hours of social science, with no particular courses in these disciplines specified. In these instances, CLEP credit may be given as "6 hrs. English Credit" or "3 hrs. Math Credit" without specifying for which English or mathematics courses credit has been awarded. To avoid possible disappointment, you should know before taking a CLEP exam what type of credit you can receive or whether you will be exempted from a required course but receive no credit.

4. *How much credit is granted for each exam you are considering, and does the college place a limit on the total amount of CLEP credit you can earn toward your degree?*

   Not all colleges that grant CLEP credit award the same amount for individual exams. Furthermore, some colleges place a limit on the total amount of credit you can earn through CLEP or other exams. Other colleges may grant you exemption but no credit toward your degree. Knowing several colleges' policies concerning these issues may help you decide which college to attend. If you think you are capable of passing a number of CLEP exams, you may want to attend a college that will allow you to earn credit for all or most of them. Check out if your institution grants CLEP policy by visiting clep.collegeboard.org/search/colleges.

5. *What is the required score for earning CLEP credit for each exam you are considering?*

   Most colleges publish the required scores for earning CLEP credit in their general catalogs or in brochures. The required score may vary from exam to exam, so find out the required score for each exam you are considering.

6. *What is the college's policy regarding prior course work in the subject in which you are considering taking a CLEP exam?*

   Some colleges will not grant credit for a CLEP exam if the candidate has already attempted a college-level course closely aligned with that exam. For example, if you successfully completed English 101 or a comparable course on another campus, you will probably not be permitted to also receive CLEP credit in that subject. Some colleges will not permit you to earn CLEP credit for a course that you failed.

7. *Does the college make additional stipulations before credit will be granted?*

   It is common practice for colleges to award CLEP credit only to their enrolled students. There are other stipulations, however, that vary from college to college. For example, does the college require you to formally apply for or to accept CLEP credit by completing and signing a form? Or does the college require you to "validate" your CLEP score by successfully completing a more advanced course in the subject? Getting answers to these and other questions will help to smooth the process of earning college credit through CLEP.

III. Preparing to Take CLEP Examinations

Test Preparation Tips

1. Familiarize yourself as much as possible with the test and the test situation before the day of the exam. It will be helpful for you to know ahead of time:

   a. how much time will be allowed for the test and whether there are timed subsections. (This information is included in the examination guides and in the CLEP Tutorial video.)

   b. what types of questions and directions appear on the exam. (See the examination guides.)

   c. how your test score will be computed.

   d. in which building and room the exam will be administered.

  e. the time of the test administration.
  f. direction, transit and parking information to the test center.
2. Register and pay your exam fee through My Account at https://clepportal.collegeboard.org/myaccount and print your registration ticket. Contact your preferred test center to schedule your appointment to test. Your test center may require an additional administration fee. Check with your test center and confirm the amount required and acceptable method of payment.
3. On the day of the exam, remember to do the following.
  a. Arrive early enough so that you can find a parking place, locate the test center, and get settled comfortably before testing begins.
  b. Bring the following with you:
    o completed registration ticket
    o any registration forms or printouts required by the test center. Make sure you have filled out all necessary paperwork in advance of your testing date.
    o a form of valid and acceptable identification. Acceptable identification must:
- Be government-issued
- Be an original document — photocopied documents are not acceptable.
- Be valid and current — expired documents (bearing expiration dates that have passed) are not acceptable, no matter how recently they may have expired
- Bear the test-taker's full name, in English language characters, exactly as it appears on the
- Registration Ticket, including the order of the names.
- Middle initials are optional and only need to match the first letter of the middle name when present on both the ticket and the identification.
- Bear a recent recognizable photograph that clearly matches the test-taker
- Include the test-taker's signature
- Be in good condition, with clearly legible text and a clearly visible photograph

Refer to the Exam Day Info page on the CLEP website (http://clep.collegeboard.org/exam-day-info) for more details on acceptable and unacceptable forms of identification.

- o military test-takers, bring your Geneva Convention Identification Card. Refer to clep.collegeboard.org/military for additional information on IDs for active duty members, spouses, and civil service civilian employees.
- o two number 2 pencils with good erasers. Mechanical pencils are prohibited in the testing room.
  c. Leave all books, papers and notes outside the test center. You will not be permitted to use your own scratch paper; it will be provided by the test center.
  d. Do not take a calculator to the exam. If a calculator is required, it will be built into the testing software and available to you on the computer. The CLEP Tutorial video will have a demonstration on how to use online calculators.
  e. Do not bring a cell phone or other electronic devices into the testing room.
4. When you enter the test room:
  a. You will be assigned to a computer testing station. If you have special needs, be sure to communicate them to the test center administrator before the day you test.
  b. Be relaxed while you are taking the exam. Read directions carefully and listen to all instructions given by the test administrator. If you don't understand the directions, ask for help before the test begins. If you must ask a question that is not related to the exam after testing has begun, raise your hand and a proctor will assist you. The proctor cannot answer questions related to the exam.
  c. Know your rights as a test-taker. You can expect to be given the full working time allowed for taking the exam and a reasonably quiet and comfortable place in which to work. If a poor testing situation is preventing you from doing your best, ask whether the situation can be remedied. If it can't, ask the test administrator to report the problem on a Center Problem Report that will be submitted with your test results. You may also wish to immediately write a letter to CLEP, P.O. Box 6656, Princeton, NJ 08541- 6656. Describe the exact circumstances as completely as you can. Be sure to include the name of the test center, the test date and the name(s) of the exam(s) you took.

Accommodations for Students with Disabilities

If you have a disability, such as a learning or physical disability, that would prevent you from taking a CLEP exam under standard conditions, you may request accommodations at your preferred test center. Contact your preferred test center well in advance of the test date to make the necessary arrangements and to find out its deadline for submission of documentation and approval of accommodations. Each test center sets its own guidelines in terms of deadlines for submission of documentation and approval of accommodations.

Accommodations that can be arranged directly with test centers include:
- ZoomText (screen magnification)
- Modifiable screen colors
- Use of a reader, amanuensis, or sign language interpreter
- Extended time
- Untimed rest breaks

If the above accommodations do not meet your needs, contact CLEP Services at clep@info.collegeboard.org for information about other accommodations.

IV. Interpreting Your Scores

CLEP score requirements for awarding credit vary from institution to institution. The College Board, however, recommends that colleges refer to the standards set by the American Council on Education (ACE). All ACE recommendations are the result of careful and periodic review by evaluation teams made up of faculty who are subject-matter experts and technical experts in testing and measurement. To determine whether you are eligible for credit for your CLEP scores, you should refer to the policy of the college you will be attending. The policy will state the score that is required to earn credit at that institution. Many colleges award credit at the score levels recommended by ACE. However, some require scores that are higher or lower than these.

Your exam score will be printed for you at the test center immediately upon completion of the examination, unless you took College Composition. For this exam, you will receive your score four to six weeks after the exam date. Your CLEP exam scores are reported only to you, unless you ask to have them sent elsewhere. If you want your scores sent to a college, employer or certifying agency, you must select this option through My Account. This service is free only if you select your score recipient at the time you register to take your exam. A fee will be charged for each score recipient you select at a later date. Your scores are kept on file for 20 years. For a fee, you can request a transcript at a later date.

The pamphlet *What Your CLEP Score Means*, which you will receive with your exam score, gives detailed information about interpreting your scores. A copy of the pamphlet is in the appendix of this Guide. A brief explanation appears below.

How CLEP Scores Are Computed

In order to reach a total score on your exam, two calculations are performed.

First, your "raw score" is calculated. This is the number of questions you answer correctly. Your raw score is increased by one point for each question you answer correctly, and no points are gained or lost when you do not answer a question or answer it incorrectly.

Second, your raw score is converted into a "scaled score" by a statistical process called *equating*. Equating maintains the consistency of standards for test scores over time by adjusting for slight differences in difficulty between test forms. This ensures that your score does not depend on the specific test form you took or how well others did on the same form. Your raw score is converted to a scaled score that ranges from 20, the lowest, to 80, the highest. The final scaled score is the score that appears on your score report.

How Essays Are Scored

The College Board arranges for college English professors to score the essays written for the College Composition exam. These carefully selected college faculty members teach at two- and four-year institutions nationwide. The faculty members receive extensive training and thoroughly review the College Board scoring policies and procedures before grading the essays. Each essay is read and scored by two professors, the sum of the two scores for each essay is combined with the multiple-choice score, and the result is reported as a scaled score between 20 and 80. Although the format of the two sections is very different, both measure skills required for expository writing. Knowledge of formal grammar, sentence structure and organizational skills are necessary for the multiple-choice section, but the emphasis in the free-response section is on writing skills rather than grammar.

Optional essays for CLEP Composition Modular and the literature examinations are evaluated and scored by the colleges that require them, rather than by the College Board. If you take an optional essay, it will be sent to the institution you designate when you take the test. If you did not designate a score recipient institution when you took an optional essay, you may still select one as long as you notify CLEP within 18 months of taking the exam. Copies of essays are not held beyond 18 months or after they have been sent to an institution.

About CLEP College Algebra

Description of the Examination

The College Algebra examination covers material that is usually taught in a one-semester college course in algebra. Nearly half of the test is made up of routine problems requiring basic algebraic skills; the remainder involves solving nonroutine problems in which candidates must demonstrate their understanding of concepts. The test includes questions on basic algebraic operations; linear and quadratic equations, inequalities, and graphs; algebraic, exponential, and logarithmic functions; and miscellaneous other topics.

It is assumed that candidates are familiar with currently taught algebraic vocabulary, symbols, and notation. The test places little emphasis on arithmetic calculations. However, an online scientific calculator (nongraphing) will be available during the examination.

The examination contains approximately 60 questions to be answered in 90 minutes. Some of these are pretest questions that will not be scored. Any time candidates spend on tutorials and providing personal information is in addition to the actual testing time.

Knowledge and Skills Required

Questions on the College Algebra examination require candidates to demonstrate the following abilities in the approximate proportions indicated.

- Solving routine, straightforward problems (about 50 percent of the examination)
- Solving nonroutine problems requiring an understanding of concepts and the application of skills and concepts (about 50 percent of the examination)

The subject matter of the College Algebra examination is drawn from the following topics. The percentages next to the main topics indicate the approximate percentage of exam questions on that topic.

Scientific Calculator

A scientific (nongraphing) calculator is integrated into the exam software, and it is available to students during the entire testing time. Students are expected to know how and when to make appropriate use of the calculator. The scientific calculator for the iBT versions of the CLEP exams, together with a brief video tutorial, is available to students as a free download for a 30-day trial period. Students and encouraged to download the calculator and become familiar with its functionality prior to taking the exam.

Students will find the online scientific calculator helpful in performing calculations (e.g, arithmetic, exponents, roots, logarithms).

25% Algebraic operations
- Factoring and expanding polynomials
- Operations with algebraic expressions
- Operations with exponents
- Properties of logarithms

25% Equations and inequalities
- Linear equations and inequalities
- Quadratic equations and inequalities
- Absolute value equations and inequalities
- Systems of equations and inequalities
- Exponential and logarithmic equations

30% Functions and their properties*
- Definition and interpretation
- Representation/modeling (graphical, numerical, symbolic, and verbal representations of functions)
- Domain and range
- Algebra of functions
- Graphs and their properties (including intercepts, symmetry, and transformations)
- Inverse functions

20% Number systems and operations
- Real numbers
- Complex numbers
- Sequences and series
- Factorials and Binomial Theorem
- Determinants of 2-by-2 matrices

*Each test may contain a variety of functions, including linear, polynomial (degree ≤ 5), rational, absolute value, power, exponential, logarithmic, and piecewise-defined.

Order of Operations

When simplifying algebraic expressions we use the following order:
1. Perform operations within a parenthesis.
2. Evaluate exponents.
3. Multiply and divide from left to right.
4. Add and subtract from left to right.

Example:

$3 + 2(4+3)^2 - 10 \div 5 = 3 + 2(7)^2 - 10 \div 5$	Perform operations in parentheses.
$= 3 + 2(49) - 10 \div 5$	Evaluate exponents.
$= 3 + 98 - 2$	Multiply and divide from left to right.
$= 99$	Add and subtract from left to right

Real numbers

The following chart shows the relationships among the subsets of the real numbers.

Real Numbers

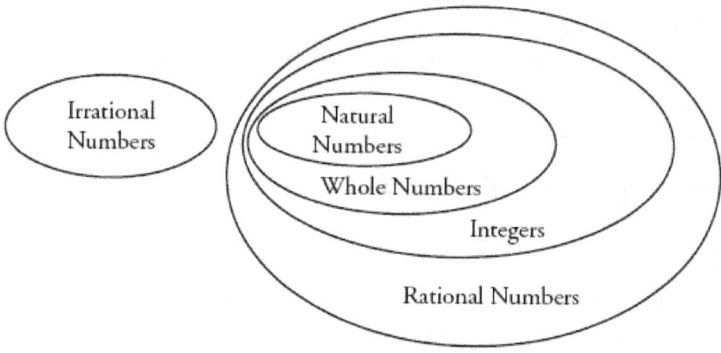

Real numbers are denoted by $\mathbb{R}$ and are numbers that can be shown by an infinite decimal representation such as 3.286275347 . . . . Real numbers include rational numbers, such as 242 and $\dfrac{23}{129}$, and irrational numbers, such as $\sqrt{2}$ and π, all of which can be represented as points along an infinite number line.

Real numbers are to be distinguished from imaginary numbers.

Real numbers are classified as follows:

Natural Numbers Denoted by $\mathbb{N}$	The counting numbers. 1, 2, 3, ...
Whole Numbers	The counting numbers along with zero. ... 0, 1, 2, 3, ...
Integers Denoted by $\mathbb{Z}$	The counting numbers, their negatives, and zero ..., $-2, -1, 0, 1, 2, ...$
Rationals, Denoted by $\mathbb{Q}$	All of the fractions that can be formed using whole numbers. Zero cannot be the denominator. In decimal form, these numbers will be either terminating or repeating decimals. Simplify square roots to determine if the number can be written as a fraction.
Irrationals	Real numbers that cannot be written as a fraction. The decimal forms of these numbers neither terminate nor repeat. Examples include π, e, and $\sqrt{2}$.

Equations and inequalities

Basic Terminology:

An equation consists of two statements linked by an equal sign (statement H1) = (statement H2)

Left Hand Side (LHS) = Right Hand Side (RHS).

If substituting a value for the variable results in LHS = RHS, or a true statement, then the value is a solution for that equation.

Example: $2x = 6$
    (LHS) (RHS)

If we substitute 3 for x, we get $2 \cdot 3 = 6$ (True).

Therefore, 3 is a solution for the equation.

Example: Is 2 a solution of $2x - 6 = 6x + 1$?

Substituting 2 for x, we get

$$2(2) - 6 = 6(2) + 1$$
$$4 - 6 = 12 + 1$$
$$-2 = 13 \text{ (False)}$$

Therefore, 2 is not a solution.

Inequalities

An inequality has the same form as an equation, but the equals sign is replaced by one of the following inequality signs:

< (less than)

> (greater than)

≤ (less than or equal to)

≥ (greater than or equal to)

The solution to an inequality is not a single value but a set of values that satisfy the inequality.

Example: $x + 2 < 7$

The solution is $x < 5$, meaning that any number less than 5 is a solution of the inequality.

Important facts about inequalities

1. Sense of an inequality: This is the direction of the inequality. The larger number is always facing the open side.

   Example: $25 > 3$ (greater than)

   Example: $3 < 25$ (less than)

2. **Notation:**

   ≥ = "Greater than or equal to".

   ≤ = "Less than or equal to".

   These relations are satisfied if either half of the relation is satisfied.

   Example: $25 \geq 3$ is true if $25 > 3$ is true or if $25 = 3$ is true. Since $25 > 3$ is true, $25 \geq 3$ is true, even though $25 = 3$ is false.

Example: $0 \leq 0$ is true if $0 < 0$ is true or $0 = 0$ is true. Since $0 = 0$ is true, $0 \leq 0$ is true, even though $0 < 0$ is false.

3. Multiplying or dividing by a negative number changes the direction of the inequality.

    Example: $-3x > 6$

    Dividing both sides by -3, we get $x < -2$ (note the change in direction)

Properties of equations and inequalities

1. We can add any real number to, or subtract any real number from, both sides of the equation (or inequality).

    Example: $3 = 3 \Rightarrow 3 + 2 = 3 + 2 \Rightarrow 5 = 5$ (still true)

    Example: $9 = 9 \Rightarrow 9 - 3 = 9 - 3 \Rightarrow 6 = 6$ (still true)

    Example: $x + 3 = 6 \Rightarrow x + 3 - 3 = 6 - 3 \Rightarrow x = 3$

2. We can multiply or divide both sides of an equation or an inequality by any real number except 0.

    Recall: When multiplying or dividing by a negative number we change the direction of the inequality.

    Example: $3 = 3 \rightarrow 3 \times 2 = 3 \times 2 \rightarrow 6 = 6$ (still true)

    Example: $8 = 8 \rightarrow \dfrac{8}{2} = \dfrac{8}{2} \rightarrow 4 = 4$ (still true)

    Example: $-2x = 6 \rightarrow \dfrac{-2x}{-2} = \dfrac{6}{-2} \rightarrow x = -3$

    Example: $6 > 2 \rightarrow 6 \times 2 > 2 \times 2 \rightarrow 12 > 4$ (still true)

    Example: $-2 < 6 \rightarrow \dfrac{-2}{-2} > \dfrac{6}{-2} \rightarrow 1 > -3$ (still true, but with reversed inequality)

    Example: $-3x \geq 5 \rightarrow \dfrac{-3x}{-3} \leq \dfrac{5}{-3} \rightarrow x \leq -\dfrac{5}{3}$ (note reversed inequality)

Matrices

A matrix is an ordered set of numbers written in rectangular form. An example matrix is shown below.

$$\begin{bmatrix} 0 & 3 & 1 \\ 4 & 2 & 3 \\ 1 & 0 & 2 \end{bmatrix}$$

Since this matrix has 3 rows and 3 columns, it is called a 3×3 matrix. If we named this matrix A, the element in the second row of the third column would be denoted as $A_{2,3}$. In general, a matrix with r rows and c columns is an $r\times c$ matrix.

Matrix addition and subtraction obey the rules of commutativity, associativity, identity, and additive inverse.

$A+B=B+A$
$A+(B+C)=(A+B)+C$
$A+0=A$
$A+0=A$

Addition, subtraction, and multiplication of matrices

Matrices can be added or subtracted only if their dimensions are the same. To add or subtract compatible matrices, simply add or subtract the corresponding elements.

Example:

$$\begin{bmatrix} 2 & 4 \\ 0 & 5 \end{bmatrix} + \begin{bmatrix} 1 & 0 \\ 7 & 8 \end{bmatrix} = \begin{bmatrix} 2+1 & 4+0 \\ 0+7 & 5+8 \end{bmatrix} = \begin{bmatrix} 3 & 4 \\ 7 & 13 \end{bmatrix}$$

Matrices can also be multiplied by a scalar. The product of a matrix and a scalar is found by multiplying each element of the matrix by the scalar.

Example: $7 \begin{bmatrix} 4 & 1 \\ 8 & 0 \end{bmatrix} = \begin{bmatrix} 28 & 7 \\ 56 & 0 \end{bmatrix}$

Determinants of matrices

Associated with every square matrix is a number called its determinant. The determinant of a matrix is typically denoted using straight brackets; thus, the determinant of matrix A is $|A|$. Use the following formula to calculate the determinant of a 2×2 matrix:

$$\begin{vmatrix} a & b \\ c & d \end{vmatrix} = ad - bc$$

Example: find the determinant of

$$\begin{bmatrix} 4 & -8 \\ 7 & 3 \end{bmatrix}$$

Use the formula for calculating the determinant.

$$\begin{vmatrix} 4 & -8 \\ 7 & 3 \end{vmatrix} = (4 \times 3) - (-8 \times 7) = 12 - (-56) = 68$$

Properties of operations

Properties are rules that apply for addition, subtraction, multiplication or division of real numbers.

1. **Commutative Property:** You can <u>change the order</u> of terms or factors as follows.
   a. <u>for addition:</u> $a + b = b + a$
   b. <u>for multiplication:</u> $ab = ba$
   (This rule does not apply for subtraction or division.)
2. **Associative Property:** You can <u>regroup the terms</u> as you like.
   a. <u>for addition:</u> $a = (b + c) = (a + b) + c$
   b. <u>for multiplication:</u> $a(bc) = (ab)c$
   (This rule does not apply for subtraction or division.)
3. **Identity Properties:** There is a number that does not change the value of another number when added to it. There is a number that does not change the value of another number when multiplied by it.
   a. <u>additive identity of 0:</u> $a + 0 = a$
   b. <u>multiplicative identity of 1:</u> $a \cdot 1 = a$

4. **Inverse Property:** For each number there is another number so that the two numbers add to 0; for each number (except 0) there is another number so that the product of the two numbers is 1.
   a. <u>for addition:</u> $a + (-a) = 0$ ($-a$ is the additive inverse of a)
   b. <u>for multiplication:</u> $a \cdot \frac{1}{a} = 1$ ($\frac{1}{a}$ is the multiplicative inverse of a, which is also called the reciprocal)
5. **Distributive Property of Multiplication Over Addition and Subtraction.** A sum or difference in parentheses can be multiplied term by term and the products added.

   <u>Example:</u> $3(x+8) = (3 \cdot x) + (3 \cdot 8) = 3x + 24$

Summary of the properties of operations

Property	of Addition	of Multiplication
Commutative	$a+b=b+a$	$ab=ba$
Associative	$a+(b+c)=(a+b)+c$	$a(bc)=(ab)c$
Identity	$a+0=a$	$a \times 1 = a$
Inverse	$a+(-a)=0$	$a \times \frac{1}{a} = 1, a \neq 0$
Distributive property of multiplication over addition and subtraction	$a(b+c)=ab+ac$	$a(b-c)=ab-ac$

Solving linear equations

1. Expand to eliminate all parentheses.
2. If there are fractions, multiply each term by the LCD to eliminate all denominators.
3. Combine terms on each side when possible.
4. Perform operations on both sides of the equation to isolate all variables on one side and all constants on the other side.

Example: solve for x: $3(x+3) = -2x+4$
$3x+9 = -2x+4$ Expand parentheses.
$3x = -2x-5$ Subtract 9 from both sides.
$5x = -5$ Add 2x to both sides.
$x = -1$ Divide both sides by 5.

Example: solve for x: $2x+9-3x+10 = 3x+x-6$
$-x+19 = 4x-6$ Combine similar terms on each side.
$-x = 4x-25$ Subtract 19 from both sides.
$-5x = -25$ Subtract 4x from both sides.
$x = 5$ Divide both sides by -5.

Example: solve for x: $3x - \dfrac{2}{3} = \dfrac{5x}{2} + 2$
$18x - 4 = 15x + 12$ Multiply each term by 6, the LCD of 2 and 3.
$18x = 15x + 16$ Add 4 to each side.
$3x = 16$ Subtract 15x from each side.
$x = \dfrac{16}{3}$ Divide each side by 3.

Solving linear inequalities

We use the same procedure used for solving linear equations, but the answer is represented in graphical form on the number line or in interval form.

Example: Solve the inequality, show its solution using interval form, and graph the solution on the number line.

$\dfrac{5x}{8} + 3 \geq 2x - 5$

$5x + 24 \geq 16x - 40$ Multiply each term by 8 to clear denominator.
$5x \geq 16x - 64$ Subtract 24 from each side.
$-11x \geq -64$ Subtract 16x from each side.

$x \leq 5\dfrac{9}{11}$ Divide each side by -11; reverse inequality sign.

Solution in interval form: $(-\infty, 5\dfrac{9}{11}]$ (Note that "[" means $5\dfrac{9}{11}$ is included in the solution.)

Graph of solution:

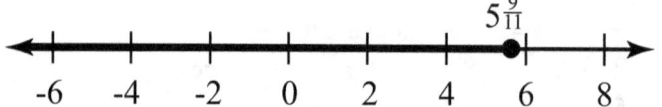

Interval and graph notation for inequalities

a. [and] mean that the lower and upper limit, respectively, are included as solutions. In graphing, a <u>closed dot</u> (•) indicates the same thing. Inclusive limits are specified with "greater than or equal to" or "less than or equal to" inequalities.

b. (and) means that the lower and upper limit, respectively, are excluded as solutions. In graphing, <u>an open</u> dot (°) indicates the same thing. Exclusive limits are specified with "greater than" or "less than" inequalities.

Example: Solve the following inequality and express your answer in both interval and graphical form.

$3x - 8 < 2(3x - 1)$

$3x - 8 < 6x - 2$ Distributive property.
$3x < 6x + 6$ Add 8 to each side.
$-3x < 6$ Subtract 6x from each side.
$x > -2$ Divide each side by -3; reverse inequality.

In graphical form:

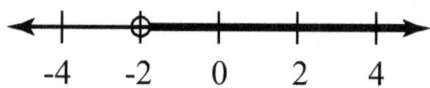

Interval form: $(-2, \infty)$ (Note that the "(" means that -2 is NOT included as a solution.)

Example: Is -2 one of the solutions of the following inequality?
$2x - 6 \leq x + 4$

substituting -2 for x, we get

$2(-2) - 6 \leq -2 + 4$

$-10 \leq 2$

This is a true statement; therefore, -2 is a solution of the inequality.

Example: Is 3 one of the solutions of the following inequality?

$$3x \leq 3 + 2$$
substituting 3 for x, we get
$$3(3) \leq 3 + 2$$
$$9 \leq 5$$

This statement is false; therefore, 3 is not a solution of the inequality.

Note: a. A linear equation has one solution, no solution, or an infinite number of solutions.
b. A linear inequality can have any number of solutions.

Graphing a linear inequality

A linear inequality in two variables is similar in form to a linear equation except that the = sign is replaced by an inequality sign of $>$, $<$, $\geq$, or $\leq$. The procedure to graph it is as follows:
1. Graph the equivalent equation with the inequality sign replaced by an equals sign. Use a solid line for this line if the inequality contains the equals sign ($\leq$ or $\geq$); use a dashed line if the inequality contains no equals sign ($<$ or $>$).
2. Pick a point on either side of the line and test whether its x- and y-values satisfy the inequality. If so, mark that region as a solution set with shading or slanted lines. If not, shade the opposite region.

Example: Identify the region that satisfies $3x + 5y < 15$.

1. We graph the equivalent equation of $3x + 5y = 15$ using a dashed line. Substituting $y = 0$ produces an x-intercept of $(5, 0)$; substituting $x = 0$ produces a y-intercept of $(0, 3)$.
2. Pick a test point on either side of this line. Pick the origin for simplicity $(0, 0)$. Substitute $x = 0$ and $y = 0$ into the inequality and test:

$0 < 15$ is true, so accept the region containing $(0, 0)$ and shade it.

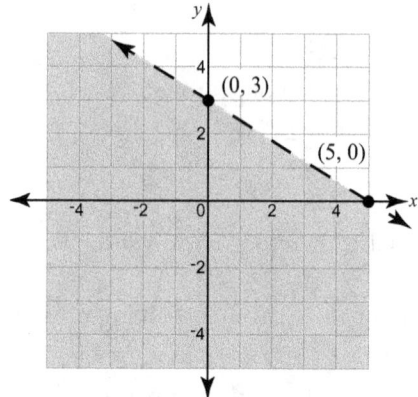

Use of "and" and "or" with inequalities
1. "And" means "intersection," the overlapping of two regions. It indicates the region that is common to the solutions of more than one inequality.
2. "Or" means "union," the joining of two regions. It indicates the region that is a solution of either inequality or both.

Example: Solve the inequalities $3x = 5y < 15$ and $y \geq 1$.
1. We found the region representing the inequality $3x + 5y < 15$ in example 2. Shade it with horizontal lines.
2. For $y \geq 1$, graph the equivalent equation of $y = 1$. This is a horizontal line through $y = 1$. Use a solid line.
3. Use the origin $(0, 0)$ as a test point. Test the inequality with $x = 0, y = 0$: $0 \geq 1$ is false. So shade the region above the line with vertical lines.
4. The solution region is the intersection region where the two regions overlap as shown.

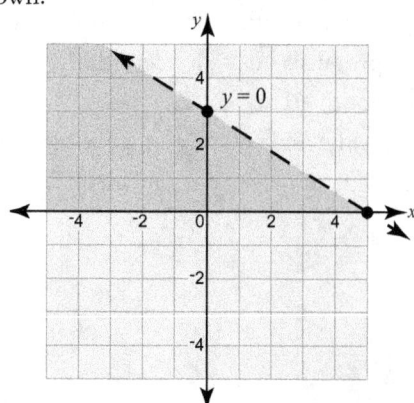

If the question had called for the solution of $3x + 5y < 15$ OR $y \geq 1$, the solution region would be as shown below.

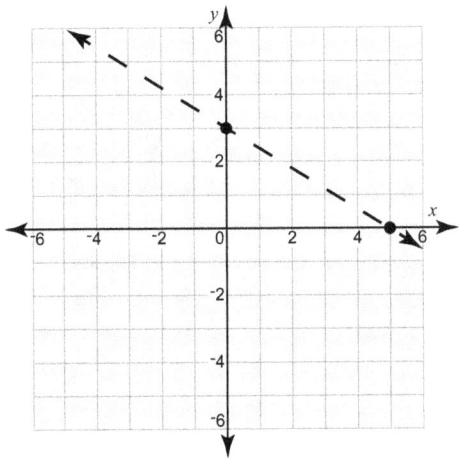

Absolute value equations

1. If a and b are real numbers, and k is a nonnegative real number, the solution of $|ax+b| = k$ is $ax+b=k$ or $ax+b=-k$

   Example: solve for x: $|2x+3| = 9$

   $$2x+3=9 \qquad\qquad 2x+3=-9$$
   $$\text{OR}$$
   $$2x+3-3=9-3 \qquad 2x+3-3=-9-3$$
   $$2x=6 \qquad\qquad 2x=-12$$
   $$x=3 \qquad\qquad x=-6$$

   Therefore, the solution is x = {3, -6}

   Example: solve for x: $|3x-1| = -3$

   Since −3 is a negative number, and an absolute value cannot be negative, there is no solution.

Absolute value inequalities

If a and b are real numbers and k is a nonnegative real number, the solution of $|ax+b| < k$ is $-k < ax+b < k$

Example: solve $|7x+3| < 25$

$-25 < (7x+3) < 25$	Rewrite original inequality.
$(-25-3) < 7x < (25-3)$	Subtract 3 from each term.
$-28 < 7x < 22$	Simplify.
$-4 < x < \dfrac{22}{7}$	Divide all terms by 7.

Solution in interval form is $\left(-4, \dfrac{22}{7}\right)$.

In graphic form:

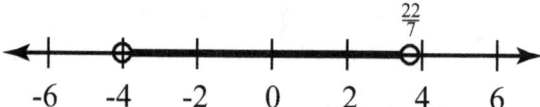

3. If a and b are real numbers and k is a nonnegative real number, the solution of $|ax+b| > k$ is $ax+b > k$ or $ax+b < -k$

Example: solve $|2x-7| > 5$

$2x-7 > 5$ $\qquad$ $2x-7 < -5$
$2x-7+7 > 5+7$ $\qquad$ $2x-7+7 < -5+7$
$2x > 12$ $\qquad$ $2x = 2$
$x > 6$ $\qquad$ $x < 1$

Solution: $x > 6$ or $x < 1$

In interval form: $(-\infty, 1) \cup (6, \infty)$

Graphically:

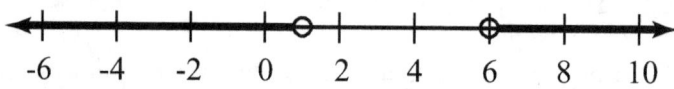

Domain and range

The domain of a function is the set of all possible inputs to the function. The range of a function is the set of all possible outputs. In some functions, both the domain and the range extend to all real numbers. Some functions have limitations on the domain, meaning that certain values are not allowed as inputs. Some functions have limitations on the range, meaning that certain values are not possible as outputs.

In the function $y = 2x + 4$, both the domain and the range extend to all real numbers. Any real number is a possible value of the input x or the output y.

In the function $y = \dfrac{1}{x-3}$, the domain includes all real numbers except 3: x cannot equal 3, because that would cause a division by zero.

In the function $y = x^2$, the range is all positive real numbers. Since the square of a real number is always positive, the possible outputs of the function do not include any negative numbers.

Functions

An equation like $y = 3x + 5$ describes a relation between the independent variable x and the dependent variable y. Thus y is written as $f(x)$, "function of x." But y is only a true function if there is a relationship between the set of all inputs or values of the independent variable (the domain) and the set of all outputs or values of the dependent variable (the range) such that each element of the domain corresponds to one element of the range. (For any input there is exactly one output.)

Example:

x	y
2	4
4	8
8	16

(This is a function.)

x	y
3	7
3	10
6	13

(This is NOT a function.)

Example: Given the function $f(x) = 3x + 5$:

Find $f(2); f(0); f(-10)$

Finding $f(2)$ means finding the function value at $x = 2$.

For $f(2); f(0); f(-10)$

Properties of functions

A relation is any set of ordered pairs. The domain of a relation is the set containing all the first coordinates of the ordered pairs, and the range of a relation is the set containing all the second coordinates of the ordered pairs.

A function is a relation in which each value in the domain corresponds to only one value in the range. It is notable, however, that a value in the range may correspond to any number of values in the domain. Thus, although a function is necessarily a relation, not all relations are functions, since a relation is not bound by this rule.

On a graph, use the vertical line test to check whether a relation is a function. If any vertical line intersects the graph of a relation in more than one point, as in the graph below, then the relation is not a function.

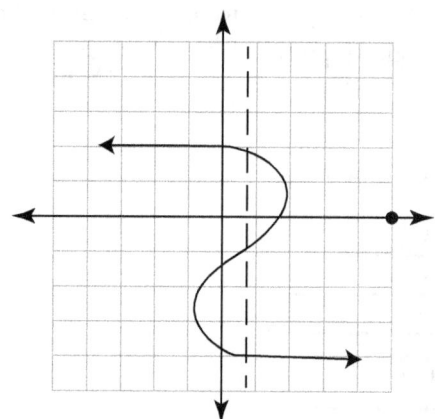

A relation is considered one-to-one if each value in the domain corresponds to only one value in the range, and each value in the range corresponds to only one value in the domain. Thus, a one-to-one relation is also a function, but it adds an additional condition.

In the same way that the graph of a relation can be examined using the vertical line test to determine whether it is a function, the horizontal line test can be used to determine if a function is a one-to-one relation. If no horizontal lines superimposed on the plot intersect the graph of the relation in more than one place, then the relation is one-to-one (assuming it also passes the vertical line test and, therefore, is a function).

As mentioned above, a function is a relation in which each value in the domain corresponds to only one value in the range. Functions can be expressed discretely, as sets of ordered pairs, or they can be expressed more generally as formulas. For instance, the function is a function that represents an infinite set of ordered pairs (x, y), where each value in the domain (x) corresponds to the same value in the range (y).

Families of functions

Some of the most commonly used function families include linear, polynomial, rational, exponential, logarithmic, and trigonometric functions. These functions, separately or in various combinations, can be used to model a range of common phenomena in finance, physics, and other fields.

Graphing a linear equation

The graph of a linear equation represents a straight line. It takes two points to define a unique straight line.

Procedure for graphing a linear equation

1. Choose only 3 values of x.
2. Substitute each chosen value of x in the equation to find the corresponding y-value.
3. Plot the 3 points and join them with a straight line.

   Note: It is typically helpful to choose the x-intercept and the y-intercept as the two key points (when possible).
   Recall:
   a. The x-intercept is the point where the line intersects the x-axis. To find this point we substitute 0 for y and solve for x.
   b. The y-intercept is the point where the line intersects the y-axis. To find this point we substitute 0 for x and solve for y.

Example: sketch the graph of the line represented by
2x + 3y = 6

Let $x = 0 \implies 2(0) + 3y = 6$
$\implies 3y = 6$
$\implies y = 2$
$\implies (0,2)$ is the y-intercept

Let $y = 0 \implies 2x + 3(0) = 6$
$\implies 2x = 6$
$\implies x = 3$
$\implies (3,0)$ is the x-intercept

Let $x = 1 \implies 2(1) + 3y = 6$
$\implies 2 + 3y = 6$ (subtract 2 from both sides)
$\implies 3y = 4$ (Divide both sides by 3)
$\implies y = \dfrac{4}{3}$
$\implies \left(1, \dfrac{4}{3}\right)$ is the third point

Plotting the 3 points or the coordinate system, we get the following graph:

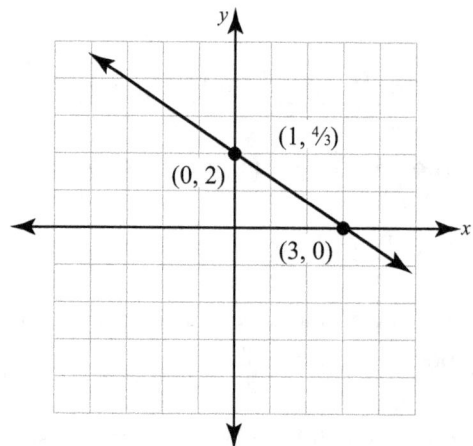

Note: Two points are sufficient to graph the line; the third point is for checking purposes.

Linear functions

A linear function can be expressed as $f(x) = mx + b$, where m and b are constants. It is called linear because it involves no quadratic or cubic variables, nor any square roots or cube roots of variables. No variables in a linear function have any exponent other than 1.

A linear function can be graphed as $y = mx + b$. The result is a straight line with slope m that intercepts the y-axis at $(0, b)$.

Example: 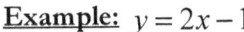 $y = 2x - 1$

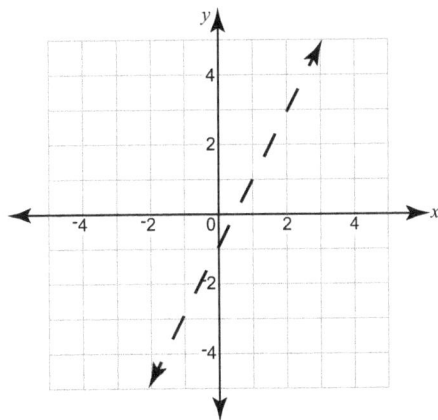

Mathematically, m could be either more or less than 0. In real-world examples of direct variation, m is almost always greater than 0, so x and y increase together. The graph of a direct variation always passes through $(0, 0)$.

Direct variation

If, in a function $y = mx + b$, b is 0, the relation is a direct variation. In a direct variation, y and x are always in the same proportion. That means that there is a constant c such that $y = cx$. Theoretically, c could be positive or negative, but in actual practice, c is almost always positive, which means that as one parameter gets larger, so does the other. The graph of a direct variation always passes through $(0, 0)$.

Example: a brand of ketchup contains 3g of sugar per ounce. This is a direct variation. If the ounces of ketchup are plotted as x and the grams of sugar as y, then $y = 3x$ as in the graph below.

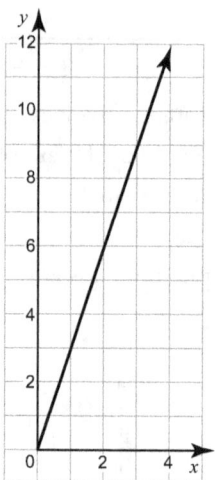

Sequences and series

Sequences and series can take on a vast range of different forms and patterns. Sequences and series are essentially two different representations of a set of numbers: a sequence is the set of numbers, and a series is the sum of the terms of the sequence. That is, a sequence such as

$a_1, a_2, a_3, ...$

has a corresponding series S such that

$S = a_1 + a_2 + a_3 + ...$

Two of the most common forms of series are the arithmetic and geometric series, both of which are discussed below.

Arithmetic series

A finite series of numbers for which the difference between successive terms is constant is called an arithmetic series. An arithmetic series with n terms can be expressed as follows, where a and d are constants. (The constant a is the first term, and d is the difference between successive terms.)

$a + (a+d) + (a+2d) + (a+3d) + ...(a+[n-1]d)$

To derive the general formula, examine the series sum for several small values of n.

n	sum
1	a
2	$2a+d$
3	$3a+3d$
4	$4a+6d$
5	$5a+10d$
6	$6a+15d$
⋮	⋮
n	$na + d\sum_{i=1}^{n-1} i$

The result in the table for n terms is found by examining the pattern of the previous series. All that is necessary, then, is to determine a closed expression for the summation.

By inspection, it can be seen that the product of n and $(n+1)$, divided by 2, is the expression for the sum of $1+2+3+4+5+\ldots+n$. Then:

$$\sum_{i=1}^{n} i = \frac{1}{2}n(n+1)$$

A simple derivation of this relationship may be made as follows:

$$S_n = 1+2+3+\ldots+n$$

Writing the terms in reverse order:

$$S_n = n+(n-1)+(n-2)+\ldots+1$$

Adding the two expressions for S_n term by term, we get

$$2S_n = (1+n)+(2+n-1)+(3+n-2)+\ldots+(n+1)$$
$$= (1+n)+(1+n)+(1+n)+\ldots+(n+1)$$
$$= n(n+1)$$

Therefore, $S_n = \dfrac{n(n+1)}{2}$

For the general case (with first term a and common difference d), therefore, the sum for a series with n terms is given by

$$na + d\sum_{i=1}^{n-1} i = na + d\frac{(n-1)(n)}{2} = \frac{1}{2}n(2a + d(n-1))$$

Often, closed formulas for series such as the arithmetic series must be found by inspection, as a more rigorous derivation is difficult. The result can be proven using mathematical induction, however.

Example: Calculate the sum of the series $1 + 5 + 9 + \ldots + 57$.

This is an arithmetic series, as the difference between successive terms, d, is constant ($d = 4$). Determine the total number of terms by subtracting the first term from the last term, dividing by d, and adding 1.

$$n = \frac{57-1}{4} + 1 = \frac{56}{4} + 1 = 14 + 1 = 15$$

That this approach works can be seen by testing simple examples. For instance, if the series is $1 + 5 + 9$, then

$$n = \frac{9-1}{4} + 1 = 3$$

There are indeed three terms in this simple series. Next, apply the formula, noting that $a = 1$.

$$\frac{1}{2}n[2a + d(n1)] = \frac{1}{2}(15)[2(1) + 4(15-1)]$$

$$= \frac{15}{2}[2 + 4(14)] = \frac{15}{2}(58) = 435$$

Thus, the answer is 435.

Geometric series

A geometric series is a series whose successive terms are related by a common factor (rather than the common difference of the arithmetic series). Assuming that a is the first term of the series and r is the common factor, the general n-term geometric series can be written as follows.

$$a + ar + ar^2 + ar^3 + \ldots + ar^{n-1}$$

The geometric series can also be written using sum notation.

$$a + ar + ar^2 + ar^3 + \ldots + ar^{n-1} = \sum_{i=0}^{n-1} ar^i$$

To derive the closed-form expression for this finite series, let the sum for n terms be defined as S_n. Multiply S_n by r.

$$S_n = a + ar + ar^2 + \ldots + ar^{n-1}$$
$$rS_n = ar + ar^2 + ar^3 + \ldots + ar^n$$

Note that if a is added to this new series, the result is the sum S_{n+1}, which has $n+1$ terms.

$$a + rS_n = a + ar + ar^2 + ar^3 + \ldots + ar^n = S_{n+1}$$

But S_{n+1} is simply $S_n + ar^n$, so the above expression can be written solely in terms of S_n.

$$a + rS_n = S_{n+1} = S_n + ar^n$$

Rearrange the result to obtain a simple formula for the geometric series.

$$a + rS_n = S_n + ar^n$$
$$a - ar^n = S_n - rS_n$$
$$a(1 - r^n) = S_n(1 - r)$$
$$S_n = a\left(\frac{1-r^n}{1-r}\right)$$

Example: find the sum of the first 5 terms of a geometric sequence whose first term is 4 and whose common factor is 3.

$a = 4, r = 3, n = 5$

$$S_n = a\left(\frac{1-r^n}{1-r}\right) = 4\left(\frac{1-3^5}{1-3}\right) = 4\frac{-242}{-2} = 4(121) = 484$$

Check: $4 + 12 + 36 + 108 + 324 = 484$ ✓

Intercepts

The intercepts of a function are the points at which the function crosses the x- or y-axis. Since the x-value of any point on the y-axis is 0, the y-intercept of any function can be found by setting x equal to 0 and using the function to find the corresponding y-value.

Example: find the y-intercept of $f(n) = x^2 + x + 4$

Let $x = 0$. $f(0) = 0^2 + 0 + 4 = 4$.

The y-intercept is at (0, 4).

Similarly, setting $f(x)$ equal to 0 and solving for x makes it possible to find an x-intercept. Whereas a function normally has one y-intercept, a function can have 0, 1, or multiple x-intercepts.

Example: find any x-intercepts for the function $f(x) = x^2 - 25$

Let $f(x) = 0$. $0 = x^2 - 25$

$x^2 = 25$

$x = \pm 5$

There are x-intercepts at (5, 0) and (–5, 0).

Systems of linear equations

A system describes a group of equations which are simultaneously true. These equations may involve one, two or more variables. A system of equations that has the same number of equations as there are variables is called a "square system". A square system will have unique solutions for each variable. A nonsquare system may have no solution at all or infinitely many solutions.

Note: An equation has no solution if the left-hand side (LHS) of the equation does not produce a true statement when compared with the right-hand side (RHS) of the equation.

Example: $x = x + 2$, which is equivalent to $0 = 2$

Note: Infinitely many solutions occur when the number of variables is less than the number of equations.

Example: When an equation is reduced to 3 = 3, it implies an infinite number of solutions.

Solving a system of equations

We will explore three basic ways of solving systems of linear equations: graphing, elimination, and substitution.

Solving a system of equations by graphing

A system of equations in two variables can be solved by solving each equation for y and graphing each equation on a common set of axes. The intersection point is the solution for x and y.

Example: solve by graphing.
$$\begin{cases} 2x - y = 1 & [1] \\ x + y = 2 & [2] \end{cases}$$
Solving [1] for y gives $y = 2x - 1$. Solving [2] for y gives $y = -x + 2$. Plot the two lines.

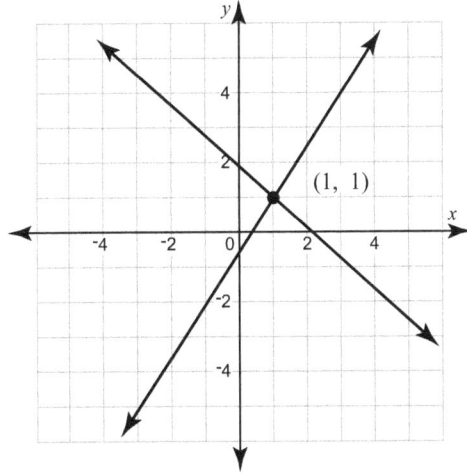

The intersection point appears to be at (1, 1), indicating a solution of $x = 1, y = 1$.

Because solving by graphing depends on the accuracy of visual inspection, solving by elimination or substitution is more reliable.

Solving a system of equations by elimination

Here we use a method which results in the elimination of one variable; this reduces the system to one equation.

Example: solve by the elimination method.
$$\begin{cases} 3x + 5y = 10 & [1] \\ 2x + 3y = 7 & [2] \end{cases}$$

Multiply $3x$ by 2 to get $6x$, multiply $2x$ by -3 to get $-6x$.

Add the new equations to eliminate x.

Eq. 1   2 $\Longrightarrow$ $6x+10y=20$ [3]

Eq. 2 -3 $\Longrightarrow$ $-6x-9y=-21$ [4]

Adding [3] and [4], we get $y=-1$ [5]

Substitute [5] into [1] and solve for x
$$3x+5(-1)=10$$
$$3x-5=10$$
$$3x=15$$
$$x=5$$

Check solutions using equation [2]:
$$2(5)+3(-1)=7$$
$$10-3=7 \checkmark$$

Solution is $x=5, y=-1$.

The answer can be represented by $(5, -1)$, since it represents the point of intersection of the two lines.

Solving a system of equations by substitution

Here we rewrite one equation in terms of a single variable. Then we substitute the expression of the variable into the second equation.

Example: Solve the system by the substitution method:
$$\begin{cases} x-y=1 & [1] \\ 3x+2y=38 & [2] \end{cases}$$

from [1], adding y to each side produces
$$x=y+1 \qquad [3]$$

substitute [3] into [2]
$$3(y+1)+2y=38$$
$$3y+3+2y=38$$
$$5y+3=38$$
$$5y=35$$
$$y=7 \quad [4]$$

Substituting [4] into 3] produces

$$x=7+1=8$$

The solution is $x=8, y=7$ or $(8,7)$.

Note: When solving a linear system of equations the elimination method is easier to use for most if not all problems.

Example: solve the system
$$\begin{cases} 2m+5n=1 & [1] \\ 6m+15n=3 & [2] \end{cases}$$
Multiply [1] by –3: $-6m-15n=-3$ [3]
Adding [3] and [2] gives $0 = 0$, which is true for any values of x. Therefore, there are infinitely many solutions. The two equations represent the same line.

Example: solve the system
$$\begin{cases} 7x+5y=25 & [1] \\ 14x+10y=-30 & [2] \end{cases}$$
Multiplying [1] by –2 gives
$$-14x-10y=-50 \quad [3]$$
Adding [2] and [3] gives $0 = -80$, which is false.

Therefore, there is no solution. The two lines are parallel and do not intersect.

Solving a system of linear inequalities

Solving systems of linear inequalities is best performed graphically. To graph a linear inequality expressed in terms of x and y, solve the inequality for y. This renders the inequality in slope-intercept form ($y = mx + b$). The point $(0, b)$ is the y-intercept, and m is the slope of the line. If the inequality is expressed only in terms of x, solve for x. When solving an inequality, remember that dividing or multiplying by a negative number will reverse the direction of the inequality sign.

If an inequality yields any of the following results in terms of y, where a is some real number, the solution set of the inequality is bounded by a *horizontal line*:

$y < a, y \leq a, y > a, \geq a$

If the inequality yields any of the following results in terms of x, then the solution set of the inequality is bounded by a *vertical line:*

$x < a, x \leq a, x > a, x \geq a$

When graphing the solution of a linear inequality, the boundary is drawn as a dashed line if the inequality sign is < or >. This indicates that points on the line do not satisfy the inequality. If the inequality sign is either ≤ or ≥, then the boundary is drawn as a solid line to indicate that the points on the line satisfy the inequality.

The line drawn as directed above is only the boundary of the solution set for an inequality. The solutions actually include the half plane bounded by the line.

Since, for any line, half of the values in the full plane (for either x or y) are greater than those defined by the line and half are less, the solution of the inequality always amounts to a half plane. Which half plane satisfies the inequality can be found by testing a point on either side of the line. The solution set can be indicated on a graph by shading the appropriate half plane.

For inequalities expressed as a function of x, shade above the line when the inequality sign is > or ≥ Shade below the line when the inequality sign is < or ≤.

For inequalities expressed as a function of y, shade to the right for > or ≥> Shade to the left for < or ≤.

The solution to a system of linear inequalities consists of the portion of the graph where the shaded half planes for all the inequalities in the system overlap. For instance, if the graph of one inequality was shaded with red and the graph of another inequality was shaded with blue, then the overlapping area would be shaded purple. The points in the purple area would be the solution set of this system.

Example: solve by graphing:

$$\begin{cases} x+y \leq 6 \\ x-2y \leq 6 \end{cases}$$

Solving the inequalities for y, they become

$$\begin{cases} y \leq -x+6 \text{ (slope } -1, y\text{-intercept } 6) \\ y \geq \frac{1}{2}x-3 \text{ (slope } \frac{1}{2}, y\text{-intercept } -3) \end{cases}$$

A graph with the appropriate shading is shown below:

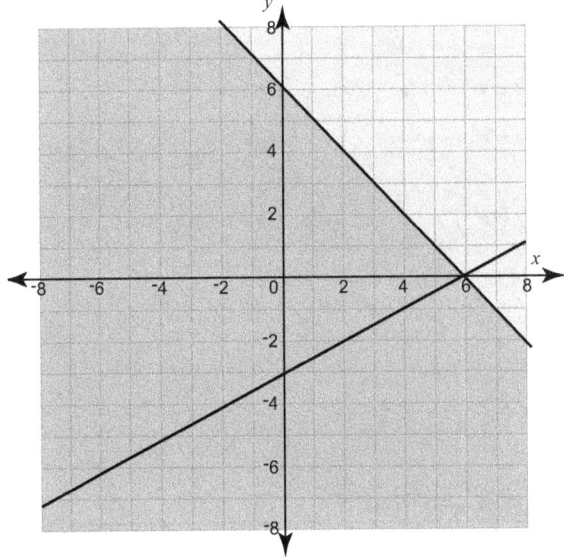

Operations with exponents

The exponent of a number indicates how many factors of that number are being multiplied together: $5^2 = 5 \times 5$, $5^3 = 5 \times 5 \times 5$. and so on. The following operations apply to numbers with exponents.

1. $a^m \times a^n = a^{m+n}$

   Example: $3^3 \times 3^2 = 3^{3+2} = 3^5 = 243$

2. $\dfrac{a^m}{a^n} = a^{m-n}$

   Example: $\dfrac{2^8}{2^5} = 2^{8-5} = 2^3 = 8$

3. $(a^m)^n = a^{mn}$

   Example: $(3^2)^3 = 3^{2 \cdot 3} = 3^6 = 729$

4. $(ab)^m = a^m b^m$

   Example: $(2 \cdot 3)^3 = (2^3)(3^3) = 216$

5. $a^{-n} = \dfrac{1}{a^n} = \left(\dfrac{1}{a}\right)^n$

   Example: $2^{-3} = \dfrac{1}{2^3} = \dfrac{1}{8}$

6. If a is any nonzero number then $a^0 = 1$.

   Example: $257^0 = 1$

7. $a^{\frac{m}{n}} = (a^{\frac{1}{n}})^m = (a^m)^{\frac{1}{n}}$

   Example: $9^{\frac{3}{2}} = (9^{\frac{1}{2}})^3 = (9^3)^{\frac{1}{2}} = 27$

8. $\sqrt[n]{a^m} = a^{\frac{m}{n}}$

   Example: $\sqrt[3]{8^2} = 8^{\frac{2}{3}} = 4$

Properties of logarithms

Exponential and logarithmic functions are complementary. The general relationship for logarithmic and exponential functions is as follows.

$y = \log_b x$ if and only if $x = b^y$

Example: $125 = 5^3$, therefore $\log_5(125) = 3$

The relationship is as follows for the exponential base e and the natural logarithm (ln).

$y = \ln x$ if and only if $e^y = x$

Example: $e^{3.5835} \approx 36$, therefore $\ln 36 \approx 3.5835$

The following properties of logarithms are helpful in solving equations.

Multiplication property

$$\log_b mn = \log_b m + \log_b n$$

Example: $\log_4(2) = 0.5, \log_4(8) = 1.5, \log_4(2 \cdot 8) = 0.5 + 1.5 = 2$

Quotient property

$$\log_b \frac{m}{n} = \log_b m - \log_b n$$

Example:
$$\log_9(243) = 2.5, \log_9(3) = 0.5, \log_9\left(\frac{243}{3}\right) = 2.5 - 0.5 = 2$$

Powers property

$$\log_b n^r = r \log_b n$$

Example: $\log_2(4) = 2, \log_2(4^3) = 3(2) = 6$

Equality property

$$\log_b n = \log_b m \text{ if and only if } n = m$$

Example: if $\log_{16}(m) = \log_{16}(64)$, $m = 64$

Change of base formula

$$\log_b n = \frac{\log_a n}{\log_a b}$$

Example: $\log_3(25) = \dfrac{\log_{10}(25)}{\log_{10}(3)}$

$$\log_b b^x = x$$

Example: $\log_6(6^3) = 3$

$$b^{\log_b x} = x$$

Example: $6^{\log_6(216)} = 216$

Solving problems involving exponential or logarithmic functions

Solving problems involving exponentials or logarithms typically involves isolating the terms containing the exponential or logarithmic function and using the inverse operation to "extract" the argument. For instance, given the following equation,

$$\ln f(x) = c$$

the function $f(x)$ can be determined by raising e to each side of the equation.

$$e^{\ln f(x)} = f(x) = e^c$$

Alternatively, if the function is in terms of an exponent e,

$$e^{f(x)} = c$$

solve by taking the natural logarithm of both sides.

$$\ln e^{f(x)} = f(x) = \ln c$$

Although these examples are in terms of e and natural logarithms, the same logic applies to exponentials and logarithms involving different bases as well.

Expanding polynomials

To multiply two binomials, use the FOIL method: First terms, Outside terms, Inside terms, Last terms.

Example: expand $(3x+1)(x-7)$.

$(3x+1)(x-7) =$
  First terms: $3x(x) +$
  Outside terms: $3x(-7) +$
  Inside terms: $1(x) +$
  Last terms: $1(-7)$
    $= 3x^2 + (-21x) + x - 7$
    $= 3x^2 - 20x - 7$

Factoring polynomials

Removing common factors

The first step in factoring any polynomial is to remove any common factor of all terms, using the distributive property.

Example: $8x^3 - 18x^2 - 4x = 2x(4x^2 - 9x - 2)$

Factoring the difference of two squares

If a polynomial has two terms, check if it is the difference of two squares. If it can be expressed as $a^2 - b^2$, it can be factored as $(a+b)(a-b)$.

Example: $9x^2 - 4 = (3x)^2 - (2)^2 = (3x+2)(3x-2)$

Factoring a trinomial in the form $x^2 + bx + c$

Look for two factors of c that add to b.

Example: factor $x^2 + 11x + 18$

Factors of 18: 1 and 18 (sum 19) ✗
               2 and 9 (sum 11) ✓

$x^2 + 11x + 18 = (x+2)(x+9)$

Factoring a trinomial in the form $x^2 - bx + c$

Look for two factors of c that add to the absolute value of b, then make both factors negative.

Example: $x^2 - 11x + 18 = (x-2)(x-9)$

Factoring a trinomial in the form $x^2 + bx - c$

Look for two factors of the absolute value of c that differ by b. Make the larger factor positive and the smaller factor negative.

Example: factor $x^2 + 5x - 24$

Factors of 24: 1 and 24 (difference 23) ✗
               2 and 12 (difference 10) ✗
               3 and 8 (difference 5) ✓

$x^2 + 5x - 24 = (x+8)(x-3)$

Factoring a trinomial in the form $x^2 - bx - c$

Look for two factors of the absolute value of c that differ by the absolute value of b. Make the larger factor negative and the smaller factor positive.

Example: $x^2 - 5x - 24 = (x-8)(x+3)$

Factoring a trinomial in the form $ax^2 + bx + c$

Look for two factors of ac that add to b.

Example: factor $4x^2 + 8x + 3$
Factors of 12: 1 and 12 (sum 13) ✗
 2 and 6 (sum 8) ✓

$4x^2 + 8x + 3 = 4x^2 + 2x + 6x + 3$ Rewrite the x-term as two terms, using the factors you have found.

$= 2x(2x+1) + 6x + 3$ Factor out the largest common factor of the first two terms.

$= 2x(2x+1) + 3(2x+1)$ Factor out the binomial in parentheses from the last two terms.

$= (2x+3)(2x+1)$ Use the distributive property to gather together the two terms of the other binomial factor.

Factoring a trinomial in the form $ax^2 - bx + c$

Example: factor $9x^2 + 3x - 2$
Look for two factors of the absolute value of ac that differ by b.
Factors of 18: 1 and 18 (difference 17) ✗
 2 and 9 (difference 7) ✗
 3 and 6 (difference 3) ✓

Make the larger factor positive and the smaller factor negative. Rewrite the polynomial, separating the x-term into terms corresponding to the two factors, positive and negative.

$9x^2 + 3x - 2 = 9x^2 + 6x - 3x - 2$ Rewrite the x-term as two terms, using the factors you have found.

$= 3x(3x+2) - 3x - 2$ Factor out the largest common factor of the first two terms.

$= 3x(3x+2) - 1(3x+2)$ Factor out the binomial in parentheses from the last two terms.

$= (3x-1)(3x+2)$ Use the distributive property to gather together the two terms of the other binomial factor.

Factoring a trinomial in the form $ax^2 - bx - c$

Look for two factors of the absolute value of ac that differ by the absolute value of b.

Make the larger factor negative and the smaller factor positive. Rewrite the polynomial, separating the x-term into terms corresponding to the two factors, positive and negative.

Example: factor $9x^2 - 3x - 2$

$9x^2 - 3x - 2 = 9x^2 - 6x + 3x - 2$ Rewrite the x-term as two terms, using the factors you have found.

$= 3x(3x - 2) + 3x - 2$ Factor out the largest common factor of the first two terms.

$= 3x(3x - 2) + 1(3x - 2)$ Factor out the binomial in parentheses from the last two terms.

$= (3x + 1)(3x - 2)$ Use the distributive property to gather together the two terms of the other binomial factor.

Factoring polynomials with four terms

Polynomials with four terms can be factored with a technique very similar to that used for trinomials:

Example: factor $8x^3 + 12x^2 - 10x - 15$

$8x^3 + 12x^2 - 10x - 15 = 4x(2x+3) - 10x - 15$ Factor out the largest common factor of the first two terms.

$= 4x^2(2x+3) - 5(2x+3)$ Factor out the binomial in parentheses from the last two terms.

$= (4x^2 - 5)(2x+3)$ Use the distributive property to gather together the two terms of the other binomial factor.

Quadratic equations and quadratic expressions

A quadratic equation is one that includes at least one squared term such as $2x^2$.

Definition: The standard form of a quadratic equation is represented by
$$ax^2 + bx + c = 0,$$
where a, b, and c are real, imaginary, or complex numbers and $a \neq 0$.

Note: Real numbers are a subset of complex numbers.

Examples: $3x^2 + 5x - 7 = 0$
$-3x^2 + 2x = 0$
$x^2 - 2 = 0$

Definition: A quadratic expression is equivalent to the left-hand side of a quadratic equation.

Examples: $2x^2 - 2x + 9$
$4x^2 - 3x$
$-2x^2 + 6$

Solving quadratic equations

Taking the square root

If it is possible to take the square root of both sides, a quadratic equation may be solved that way.

Example:
$4x^2 + 4x + 1 = 49$

$(2x+1)^2 = 7^2$ Represent both sides as squares.

	$2x+1 = \pm 7$	Take square root of each side.
$2x+1=7$	$2x+1=-7$	Find two solutions.
$2x=6$	$2x=-8$	Subtract 1 from each side.
$x=3$	$x=-4$	Divide each side by 2.
	$x=3,-4$	Combine solutions.

Solving quadratic equations by factoring

If the left-hand side of a quadratic equation in standard form can be factored, the equation can be solved by setting each factor to 0.

Example: solve for x: $x^2 - x - 6 = 0$

	$(x+2)(x-3)=0$	Equation is true if either factor equals 0.
$x+2=0$	$x-3=0$	Solve by setting each factor equal to 0.
$x=-2$	$x=3$	Find two solutions.
	$x=-2,3$	Combine solutions.

In general, if $x = a$ is a solution of a quadratic equation in standard form, then $x - a$ is a factor of the expression on the left-hand side of the equation, and vice versa.

Example: 3 is a solution of $x^2 + 2x - 15 = 0$, since $3^2 + 2(3) - 15 = 9 + 6 - 15 = 0$. Therefore, $(x-3)$ is a factor of $x^2 + 2x - 15$: $x^2 + 2x - 15 = (x-3)(x+5)$

Solving quadratic equations with the quadratic formula

Quadratic formula: When a quadratic equation is not factorable we use the quadratic formula, which <u>always yields</u> a solution. The solution of $ax^2 + bx + c = 0$ is given by the formula:

$$x = \frac{-b \pm \sqrt{b^2 - 4ac}}{2a}$$

This formula yields two solutions:

$$x_1 = \frac{-b + \sqrt{b^2 - 4ac}}{2a}, x_2 = \frac{-b - \sqrt{b^2 - 4ac}}{2a}$$

Solving quadratic equations

Example: Solve for x: $3x^2 + 5x - 3 = 0$

$a = 3; b = 5; c = -3$

$$x = \frac{-b \pm \sqrt{b^2 - 4ac}}{2a}$$

$$= \frac{-5 \pm \sqrt{5^2 - 4(3)(-3)}}{2(3)}$$

$$= \frac{-5 \pm \sqrt{61}}{6}$$

$$x_1 = \frac{-5 + \sqrt{61}}{6}, \quad x_2 = \frac{-5 - \sqrt{61}}{6}$$

Quadratic equations and imaginary numbers

Sometimes, when solving a quadratic equation, you need to take the square root of a negative number. If so, the equation has no real-number solutions: since the square of either a positive or negative number is positive, no real number could be the square root of a negative number. To deal with the square roots of negative numbers, mathematicians have created the concept of i, the square root of –1. The square roots of other negative numbers can be given in terms of i, using standard operations with radicals:

$$\sqrt{-25} = \sqrt{(25)(-1)} = \sqrt{25}\left(\sqrt{-1}\right) = 5i \cdot$$

Example: solve for x: $x^2 + 36 = 0$

$$x^2 = -36$$

$$x = \pm 6i$$

$$x_1 = 6i, x_2 = -6i$$

Using the quadratic formula often yields a solution consisting of a real number plus or minus an imaginary number. Such numbers are called complex numbers.

Example: solve for x: $2x^2 - 4x + 4 = 0$

$$x = \frac{-b \pm \sqrt{b^2 - 4ac}}{2a}$$
$$= \frac{-(-4) \pm \sqrt{(-4)^2 - 4(2)(4)}}{2(2)}$$
$$= \frac{4 \pm \sqrt{-16}}{4}$$
$$= \frac{4 \pm 4i}{4}$$
$$= 1 \pm i$$
$$x_1 = 1+i, x_2 = 1-i$$

Complex numbers

The set of complex numbers is denoted by $\mathbb{C}$. The set $\mathbb{C}$ is defined as $\{a+bi : a,b \in \mathbb{R}\}$ ("$\in$" means "element of"). In other words, complex numbers are an extension of real numbers made by attaching an imaginary number i, which satisfies the equality $i^2 = -1$. Complex numbers are of the form $a+bi$, where a and b are real numbers and $i = \sqrt{-1}$. Thus, a is the real part of the number and b is the imaginary part of the number. When i appears in a fraction, the fraction is usually simplified so that i is not in the denominator. The set of complex numbers includes the set of real numbers, where any real number n can be written in its equivalent complex form as $n+0i$. In other words, it can be said that $\mathbb{R} \subseteq \mathbb{C}$ (or $\mathbb{R}$ is a subset of $\mathbb{C}$).

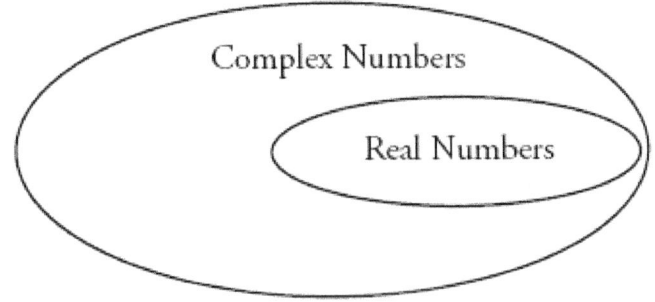

The number $3i$ has a real part 0 and imaginary part 3; the number 4 has a real part 4 and an imaginary part 0. As another way of writing complex numbers, we can express them as ordered pairs:

Complex number	Ordered pair
$3+2i$	$(3, 2)$
$\sqrt{3}+\sqrt{3}i$	$(\sqrt{3},\sqrt{3})$
$7i$	$(0, 7)$
$\dfrac{6+2i}{7}$	$\left(\dfrac{6}{7},\dfrac{2}{7}\right)$

The basic operations for complex numbers can be summarized as follows, where $z_1 = a_1 + b_1 i$ and $z_2 = a_2 + b_2 i$. Note that the operations are performed in the standard manner, i being treated as a standard radical value. The result of each operation is written in the standard form for complex numbers. Also note that the complex conjugate of a complex number $z = a + bi$ is denoted as $z^* = a - bi$.

$$z_1 + z_2 = (a_1 + b_1) + (b_1 + b_2)i$$
$$z_1 - z_2 = (a_1 - a_2) + (b_1 - b_2)$$
$$z_1 z_2 = (a_1 a_2 - b_1 b_2) + (a_1 b_2 - a_2 b_1)i$$
$$\frac{z_1}{z_2} = \frac{z_1 z_2^*}{z_2 z_2^*} = \frac{a_1 a_2 + b_1 b_2}{a_2^2 + b_2^2} + \frac{a_2 b_1 + a_1 b_2}{a_2^2 + b_2^2}i$$

The rational root theorem

The rational root theorem, also known as the rational zero theorem, allows determination of all possible rational roots (or zeroes) of a polynomial equation with integer coefficients. (A root is a value of x such that $P(x) = 0$.) Every rational root of $P(x)$ can be written as $x = \dfrac{p}{q}$, where p is an integer factor of the constant term a_0 and q is an integer factor of the leading coefficient a_n.

Example: find the possible rational roots of $3x^3 - 7x^2 + 3x - 2$.

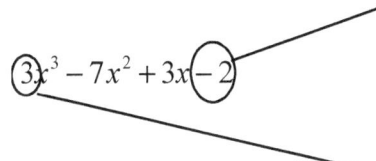

p must be an integer factor of −2: 1, −1, 2, or −2.

q must be an integer factor of 3: 1 or 3.

$$\frac{p}{q} = \frac{1, -1, 2, \text{ or} -2}{1 \text{ or } 3} = 1, -1, 2, -2, \frac{1}{3}, -\frac{1}{3}, \frac{2}{3}, \text{ or } -\frac{2}{3}$$

The rational root theorem guarantees that any rational roots of $3x^3 - 7x^2 + 3x - 2$ will be in the list just given, but it does NOT guarantee that every item in the list will be a root. Test each possibility. The only result that works is $x = 2$.

The complex conjugate root theorem

For a polynomial $P(x)$ with real coefficients, if $P(x)$ has a complex root $a + bi$, then it must also have a complex root $a - bi$. In the case of quadratic equations, this becomes obvious when the quadratic formula is used, since the imaginary part of any root is always calculated as "plus or minus." However, it also holds for higher-power polynomials.

Quadratic inequalities

To solve a quadratic inequality, first gather all terms on the left side of the inequality sign, leaving zero on the right.

Example: $x^2 < 5x + 6 \rightarrow x^2 - 5x - 6 < 0$

Then replace the sign of the inequality with an equals sign and solve the resulting quadratic equation by factoring or by using the quadratic formula:

$$x^2 - 5x - 6 = 0 \rightarrow (x+1)(x-6) = 0 \rightarrow x = -1, 6$$

The last step is to restore the inequality signs. This can be done by graphing the quadratic equation as a function. In this example, since the first term of the equation is positive, the resulting graph will be a parabola opening upward, with zeros at −1 and 6.

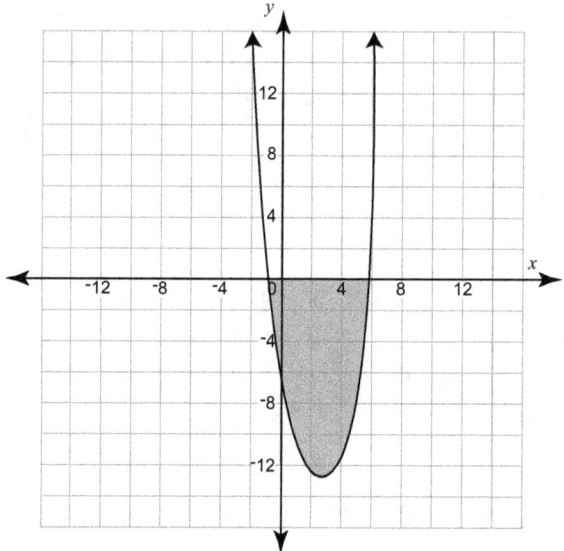

To restore the inequality signs, note that in this example, the solutions are supposed to be less than zero. Those solutions will be found in the part of the parabola that goes below the x-axis, namely the section between −1 and 6. Therefore, the solutions are $-1 < x < 6$.

Modeling functions

Functions can be represented in a variety of ways, including as a symbolic expression (for instance, $f(x) = 3x^2 - \sin x$), a graph, a table of values, and a common-language expression (for example, "the speed of the car increases linearly from 0 to 100 miles per hour in 12 seconds"). The ability to convert among various representations of a function depends on how much information is provided. For instance, although a graph of a function can provide some clues as to its symbolic representation, it is often difficult or impossible to obtain an exact symbolic form based only on a graph. The same difficulty applies to tables.

Converting from a symbolic form to a graph or table, however, is relatively simple, especially if a computer is available. The symbolic expression need simply be evaluated for a representative set of points that can be used to produce a sufficiently detailed graph or table. For example, the equation

$y = 9x$ describes the relationship between y, the total number of dollars earned, and x, the number of $9 sunglasses sold. In a relationship of this type, one of the quantities (e.g., total amount earned) is dependent on the other (e.g., number of sunglasses sold). These variables are known as the dependent and independent variables, respectively. A table using this data would appear as:

number of sunglasses sold	1	5	10	15
total dollars earned	9	45	90	135

Each (x, y) relationship between a pair of values is called a coordinate pair and can be plotted on a graph. The coordinate pairs (1, 9), (5, 45), (10, 90), and (15, 135) are plotted on the graph below.

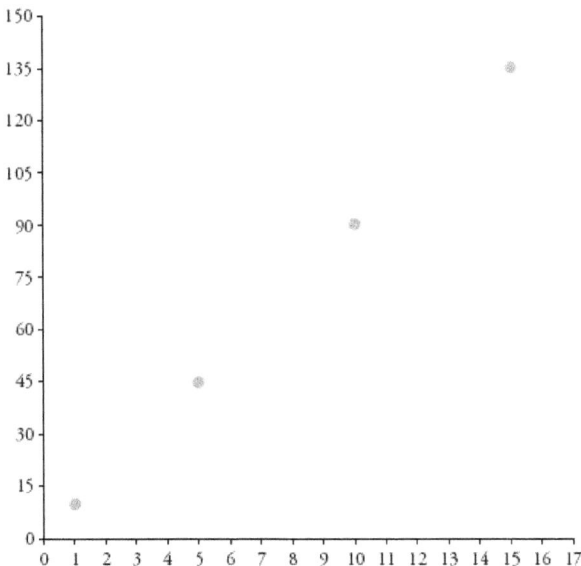

The graph shows a linear relationship. A linear relationship is one in which the change in two quantities is in a constant proportion. Doubling the change in x also doubles the change in y. On a graph, a straight line depicts a linear relationship.

Modeling functions 63

The function or relationship between two quantities may be analyzed to determine how one quantity depends on the other. For example, the function $y = 2x + 1$ below shows a relationship between y and x.

The relationship between two or more variables can be analyzed using a table, graph, written description or symbolic rule. The function $y = 2x + 1$ is written as a symbolic rule. The same relationship is also shown in the table below:

x	0	2	3	6	9
y	1	5	7	13	19

A relationship could be written in words by saying the value of y is equal to two times the value of x, plus one. This relationship could be shown on a graph by plotting given points such as the ones shown in the table above.

Another way to describe a function is as a process in which one or more numbers are input into an imaginary machine that produces another number as the output. If 3 is input as x into a machine with a process of 2x +1, the output, y, will equal 7.

In real situations, relationships can be described mathematically. The function $y = x + 1$ can be used to describe the idea that people age one year on their birthday. To describe the relationship in which a person's monthly medical costs are 6 times a person's age, we could write $y = 6x$. The monthly cost of medical care could be predicted using this function. A 20-year-old person would spend $120 per month (120 = 20 × 6). An 80-year-old person would spend $480 per month (480 = 80 × 6). Therefore, one could analyze the relationship to say: as you get older, monthly medical costs increase $6.00 each year.

Example: What is the equation that expresses the relationship between x and y in the table below?

x	0	1	2	3	4	5
y	3	5	7	9	11	13

We will write the equation as $y = mx + b$. We can determine m (the slope) as the change in y between any two points divided by the change in x between any two points. In this table, the change in y between two points is 2 and the change in x between two points is 1, so the slope m is $\frac{2}{1} = 2$. Find b by setting x to zero. In this table, when $x = 0$, $y = 3$, so $b = 3$, and the equation is $y = 2x + 3$.

Finding powers of a binomial (the binomial theorem)

The figure below is known as Pascal's triangle. Only a portion of it is shown here. The left and right borders are all 1's. In the interior, each number is the sum of the two numbers above it to left and right.

Row 0:   1

Row 1:   1   1

Row 2:   1   2   1

Row 3:   1   3   3   1

Row 4:   1   4   6   4   1

Row 5:   1   5   10   10   5   1

Row 6:   1   6   15   20   15   6   1

Pascal's triangle can be used to find powers of a binomial without repeatedly multiplying. Notice that each row has a number, starting with row 0. The numbers in each row are the number coefficients of the terms of ($a + b$) raised to the power of the row number. The variables in the terms start with a^n, followed by $a^{n-1}b$, $a^{n-2}b^2$, and so on, with decreasing exponents of a and increasing exponents of b, ending with ab^{n-1} and b^n. Putting the coefficients from row n together with the variable exponents in the series produces the binomial expansion of $(a+b)^n$. For example: $(a+b)^3 = a^3 + 3a^2b + 3ab^2 + b^3$

Any binomial can be expanded in this fashion by setting its two terms equal to a and b.

Example: Find $(2x-3)^4$.

$a = 2x, b = -3$

$(a+b)^4 = a^4 + 4a^3b + 6a^2b^2 + 4ab^3 + b^4$

$= (2x)^4 + 4(2x)^3(-3) + 6(2x)^2(-3)^2 + 4(2x)(-3)^3 + (-3)^4$

$= 16x^4 - 96x^3 + 216x^2 - 216x + 81$

Composition of functions

Composition of functions is a way of combining functions such that the range of one function is the domain of another. For instance, the composition of functions f and g can be either $f \circ g$ (the composite of f with g) or $g \circ f$ $g + f$ (the composite of g with f). Another way of writing these compositions is f $f(g(x))$ and $g(f(x))$. The domain of the composition includes all values x such that $g(x)$ is in the domain of $f(x)$.

> **Example:** What is the composition $f \circ g$ for the functions $f(x) = ax$ and $g(x) = bx^2$?
>
> The correct answer can be found by substituting the function $g(x)$ into $f(x)$:
> $$f(g(x)) = a \cdot g(x) = abx^2$$
>
> On the other hand, the composition $g \circ f$ would yield a different answer.
> $$g(f(x)) = b \cdot (f(x))^2 = b(ax)^2 = a^2bx^2$$

Inverses of functions

The inverse of a function $f(x)$ is typically labeled $f^{-1}(x)$ and satisfies the following two relations:

$$f(f^{-1}(x)) = x$$
$$f^{-1}(f(x)) = x$$

For a function $f(x)$ to have an inverse, it must be one-to-one. This fact is easily seen, since both $f(x)$ and $f^{-1}(x)$ must satisfy the vertical line test (that is, both must be functions). A function takes each value in a domain and relates it to only one value in the range. Logically, then, the inverse must do the same, only backwards: relate each value in the range to a single value in the domain.

Finding inverses of functions

Finding the inverse of a function can be a difficult or impossible task, but there are some simple approaches that can be followed in many cases. The simplest method for finding the inverse of a function is to interchange the variable

and the function symbols and then solve to find the inverse. The approach is summarized in the outline below, given a one-to-one function $f(x)$.
1. Replace the symbol $f(x)$ with x
2. Replace all instances of x in the function definition with $f^{-1}(x)$ (or y or some other symbol)
3. Solve for $f^{-1}(x)$.
4. Check the result using $f(f^{-1}(x)) = x$ or $f^{-1}(f(x)) = x$.

Example: Determine if the function $f(x) = x^2$ has an inverse. If so, find the inverse.

First, determine if $f(x)$ is one-to-one. Note that $f(1) = f(-1) = 1$, so $f(x)$ is not one-to-one and therefore has no inverse function.

Example: Determine if the function $f(x) = x^3 + 1$ has an inverse. If so, find the inverse.

The function f (x) 5 x 3 1 1 has an inverse because it increases monotonically for $x > 0$ and decreases monotonically for $x < 0$. As a result, it is one-to-one, and the inverse exists. To calculate the inverse, let y be $f^{-1}(x)$. Replace $f(x)$ with x and replace x with y.

$$f(x) = x^3 + 1 \to x = y^3 + 1$$

Solve for y.

$$x - 1 = y^3$$
$$y = \sqrt[3]{x-1}$$
$$f^{-1}(x) = \sqrt[3]{x-1}$$

Test the result.
$$f^{-1}(f(x)) = \sqrt[3]{(x^3+1)-1}$$
$$= \sqrt[3]{x^3+1-1} = \sqrt[3]{x^3} = x$$
The result is thus correct.

Operations with radicals

Radicals are inverse operators of exponents: $\sqrt[n]{a}$ is called the "nth root of a" and means the number that would have to be multiplied by itself n times to produce a: $x = \sqrt[2]{a}$ means that $x \cdot x = a$, $x = \sqrt[3]{a}$ means $x \cdot x \cdot x = a$, and so on. The number under the radical sign, in this case a, is called the radicand. The number above the radicand to the left is called the index or root. When the root is omitted, it is always assumed to be 2. That is, $\sqrt{x} = \sqrt[2]{x}$.

Every positive number has two square roots, one positive and one negative. The square root of 16, for instance, is either 4 or –4, since (+4)(+4) = 16 and (–4)(–4) = 16. We can write the two results together as +4 is called the principal square root of 16. In many problems, the principal square root is the only answer that makes sense.

Example: find the length of one side of a square room having an area of 16 square feet. Here the only answer is +4 ft, since a length of –4 ft is meaningless.

Addition and subtraction of radicals

1. We can only add or subtract radicals that have the same index and the same radicand.

   Example: $5\sqrt[3]{2} - 3\sqrt[3]{2} = 2\sqrt[3]{2}$

2. If the radicand is raised to a power equal to the index, the root operation cancels out the power operation.

   Example: $\sqrt[5]{7^5} = 7$

3. If the radicand is raised to a power <u>different</u> from the index, convert the radical to its exponential form and apply laws of exponents.

   Example: $\sqrt[3]{a^4} = (a^4)^{\frac{1}{3}} = a^{\frac{4}{3}}$

Multiplication and division of radicals

1. <u>Multiplication:</u> If the indexes or roots are the same, just multiply the radicands and keep the same index.

   Example: $\sqrt{3} \times \sqrt{8} = \sqrt{3 \times 8} = \sqrt{24}$

If the indexes or roots are not the same but the radicands are the same, convert each number to its exponent form and apply laws of exponents.

Example: $\sqrt{a}\left(\sqrt[3]{a}\right) = a^{\frac{1}{2}}a^{\frac{1}{3}} = a^{\left(\frac{1}{2}+\frac{1}{3}\right)} = a^{\frac{5}{6}}$

2. Division: to divide by a radical denominator, we must eliminate the radical from the denominator. We call this operation "rationalizing" the denominator.

   case 1: If the denominator has a single square root, we multiply both the numerator and denominator by the denominator.

   Example: $\dfrac{3}{\sqrt{2}} = \dfrac{3}{\sqrt{2}}\left(\dfrac{\sqrt{2}}{\sqrt{2}}\right) = \dfrac{3\sqrt{2}}{2}$

   case 2: If the denominator has two terms, we multiply the denominator and the numerator by the conjugate of the denominator. The conjugate is produced by changing the sign between the two terms from plus to minus or from minus to plus.

   Example:

   $\dfrac{3}{5+\sqrt{2}} = \dfrac{3(5-\sqrt{2})}{(5+\sqrt{2})(5-\sqrt{2})}$ Multiply numerator and denominator by the conjugate of the denominator.

   $= \dfrac{3(5-\sqrt{2})}{5^2 - (\sqrt{2})^2}$ Difference of two squares.

   $= \dfrac{15 - 3\sqrt{2}}{23}$ Simplify using distributive property.

   In this example, the conjugate of $5+\sqrt{2}$ is $5-\sqrt{2}$.

Transformational geometry: translations, rotations, reflections, and scaling

Transformational geometry is the study of the manipulation of objects through movement, rotation, and scaling. The transformed version of an object is called its image. If the original object is labeled with letters, such as ABCD, the image can be labeled with the same letters followed by a prime symbol: A'B'C'D'.

Transformations can be characterized in different ways.

Types of transformations

An isometry is a linear transformation that maintains the dimensions of a geometric figure.

Symmetry is exact similarity between two parts or halves, as if one were a mirror image of the other.

A translation is a transformation that "slides" an object a fixed distance in a given direction. The original object and its translation have the same shape and size, and they face in the same direction.

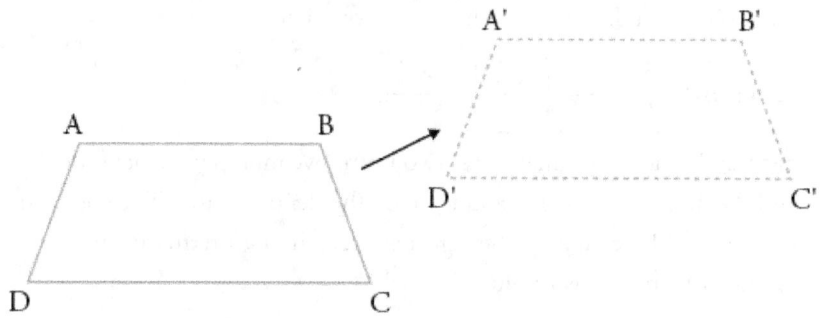

A rotation is a transformation that turns a figure about a fixed point, which is called the center of rotation. An object and its rotation are the same shape and size, but the figures may be oriented in different directions. Rotations can occur in either a clockwise or a counterclockwise direction.

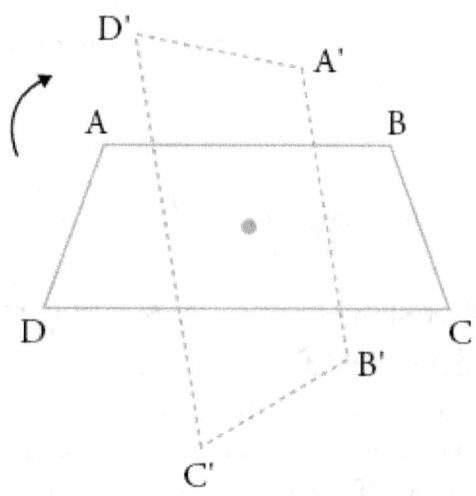

An object and its reflection have the same shape and size, but the figures face in opposite directions. The line (where a hypothetical mirror may be placed) is called the line of reflection. The distance from a point to the line of reflection is the same as the distance from the point's image to the line of reflection.

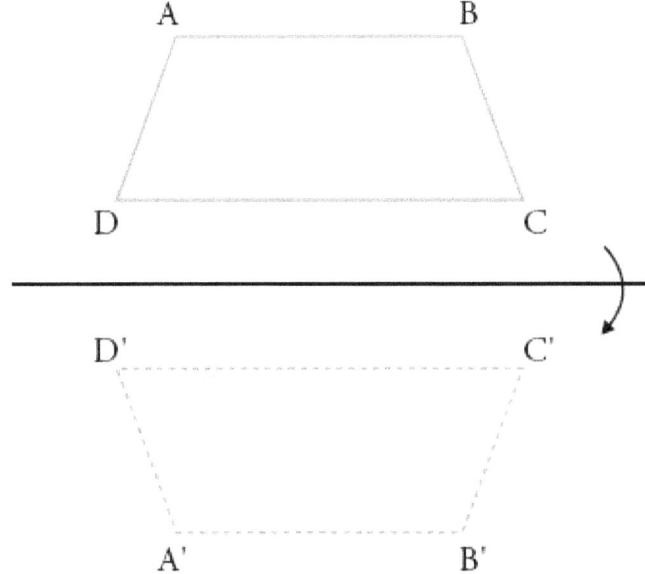

The examples of a translation, a rotation, and a reflection given above are for polygons, but the same principles apply to the simpler geometric elements of points and lines. In fact, a transformation performed on a polygon can be viewed equivalently as the same transformation performed on the set of points (vertices) and lines (sides) that compose the polygon. Thus, to perform complicated transformations on a figure, it is helpful to perform the transformations on all the points (or vertices) of the figure, then reconnect the points with lines as appropriate.

Multiple transformations can be performed on a geometrical figure. The order of these transformations may or may not be important. For instance, multiple translations can be performed in any order, as can multiple rotations (around a single fixed point) or reflections (across a single fixed line). The order of the transformations becomes important when several types of transformations are performed or when the point of rotation or the line of reflection changes among transformations. For example, consider a translation of a given distance upward and a clockwise rotation by 90° around a fixed point. Changing the order of these transformations changes the result.

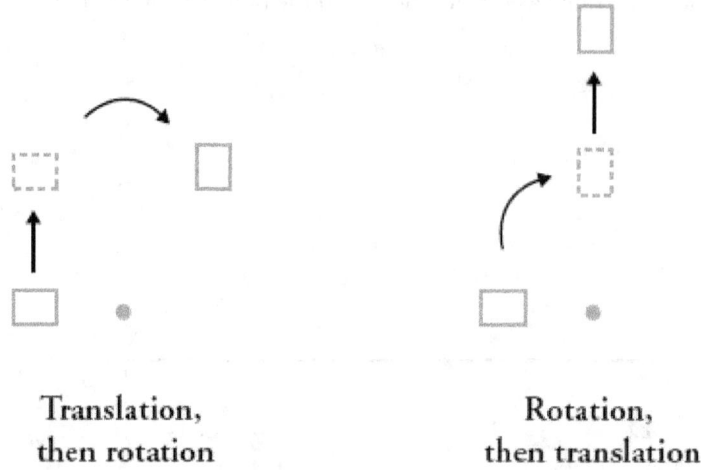

Translation, then rotation

Rotation, then translation

As shown, the final position of the box is different, depending on the order of the transformations. Thus, it is crucial that the proper order of transformations (whether determined by the details of the problem or some other consideration) be followed.

Example: Find the final location of a point at (1, 1) that undergoes the following transformations: rotate 90° counterclockwise about the origin; translate distance 2 in the negative y-direction; reflect about the y-axis.

First, draw a graph of the x- and y-axes and plot the point at (1, 1).

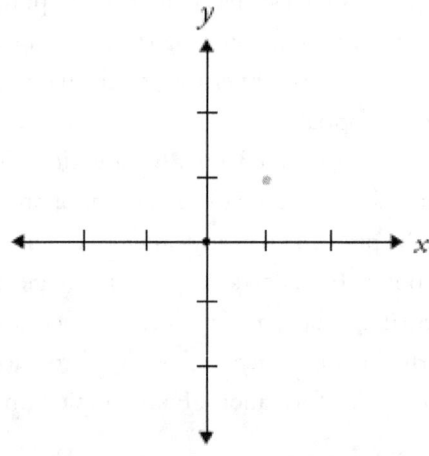

Next, perform the rotation. The center of rotation is the origin, and the rotation is in the counterclockwise direction. In this case, the even value of 90° makes the rotation simple to do by inspection. Next, perform a translation of distance 2 in the negative y direction (down). The results of these transformations are shown below.

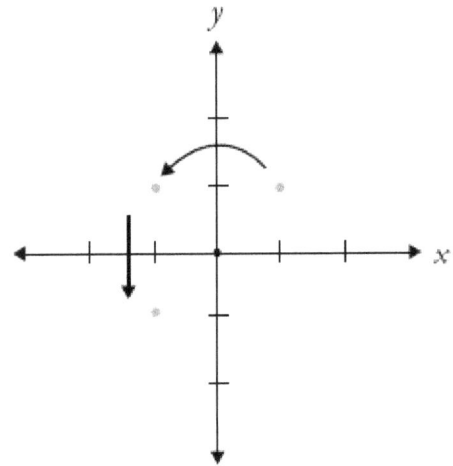

Finally, perform the reflection about the y-axis. The final result, shown below, is a point at (1, −1).

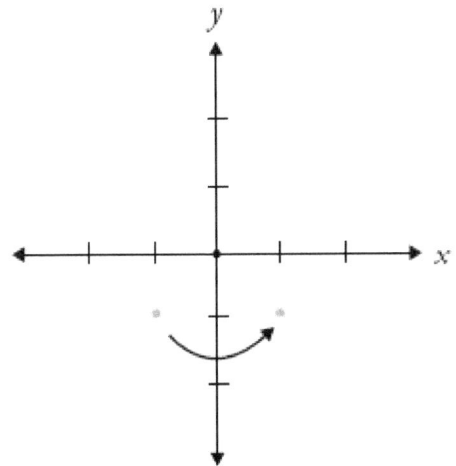

Using this approach, polygons can be transformed on a point-by-point basis. For some problems, there is no need to work with coordinate axes. For instance, the problem may simply require transformations without respect to any absolute positioning.

Example: Rotate the following regular pentagon by 36° about its center, and then reflect it about a horizontal line.

First, perform the rotation. In this case, the direction is not important because the pentagon is symmetric. As it turns out in this case, a rotation of 36° yields the same result as flipping the pentagon vertically (assuming the vertices of the pentagon are indistinguishable).

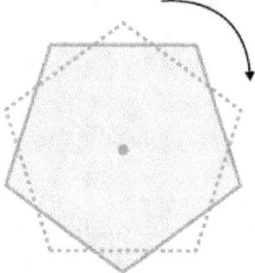

Finally, perform the reflection. Note that the result here is the same as a downward translation (assuming the vertices of the pentagon are indistinguishable).

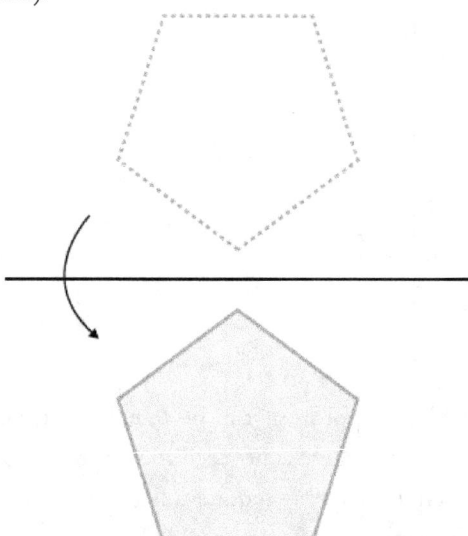

Symmetries

Symmetries in a function can also be described in terms of reflections or "mirror images." A function can be symmetric about the the y-axis (but not about the x-axis, except for the function $f(x)=0$, since every function must pass the vertical line test). A function is symmetric about the y-axis if for every point (x,y) that is included on the graph of the function, the point $(-x,y)$ is also included on the graph. Consider the function $f(x)=x^2$. Note that for each point (x,x^2) on the graph, the point $(-x,x^2)$ is also on the graph. The symmetry of the function about the y-axis can also be seen in the graph below.

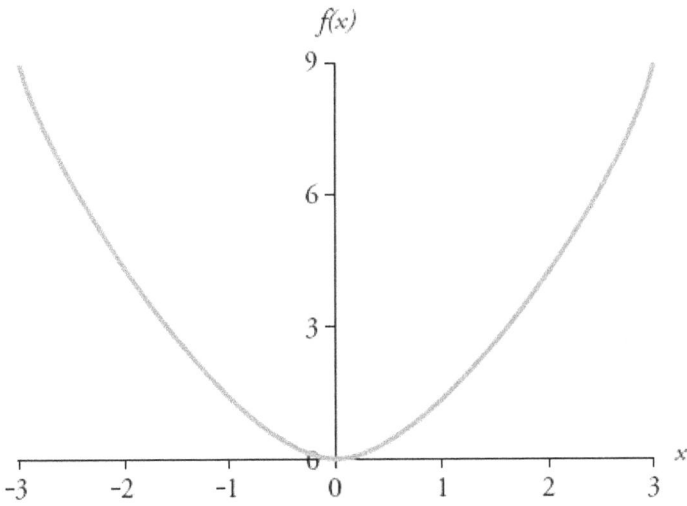

A function that is symmetric about the y-axis is also called an even function. Although functions cannot be symmetric about the x-axis, relations that do not obey the vertical line test can be symmetric in this way. A relation is symmetric about the x-axis if for every point (x,y) in the graph of the relation, the point $(x,-y)$ is also in the graph.

Consider, for instance, the relation Ï $g(x)=\pm\sqrt{x}$. For every value of x in the domain, the points $(x,\sqrt{x})$ and $(x,-\sqrt{x})$ are both in the graph, as shown below.

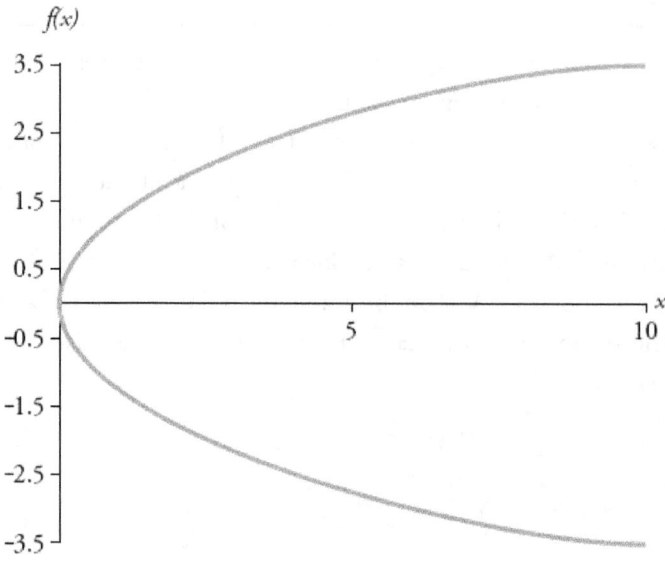

Functions may also be symmetric with respect to the origin. Such functions are called odd (or antisymmetric) functions and are defined by the property that for any point (x, y) on the graph of the function, the point is also on the graph of the function. The function $(-x, -y)$ for instance, is symmetric with respect to the origin, as shown in the graph below.

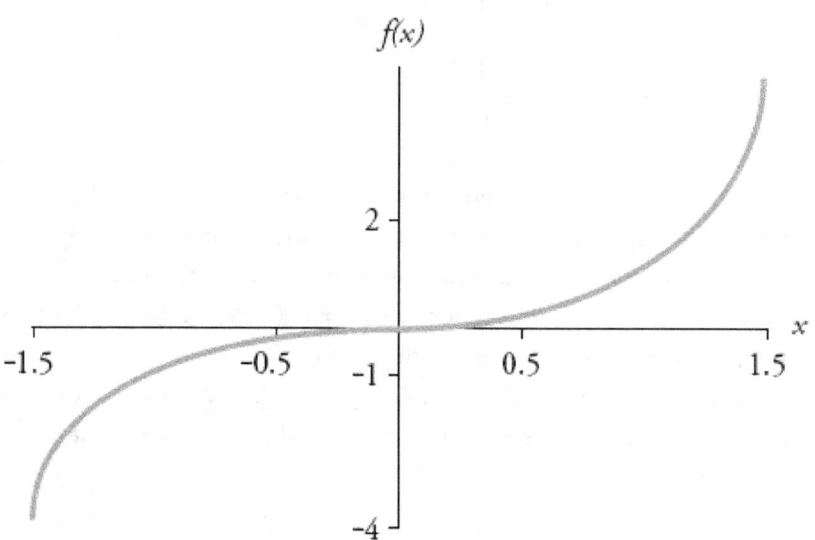

Piecewise functions

In a piecewise function, different functions are specified for different intervals of the domain.

Example: $f(x) = \begin{cases} -2x & \text{if } x < 2 \\ 0.5x+1 & \text{if } x \geq 2 \end{cases}$

Piecewise functions can be continuous or discontinuous. A piecewise function is continuous for a certain interval if it is defined for every point in the interval and if it can be drawn with a single stroke.

A piecewise function is discontinuous during a certain interval if it is undefined at any point or if there is a jump in value between subintervals. The function above is discontinuous, though it is defined for every point in the domain of real numbers, because the value jumps from –4 to 2 at $x = 2$.

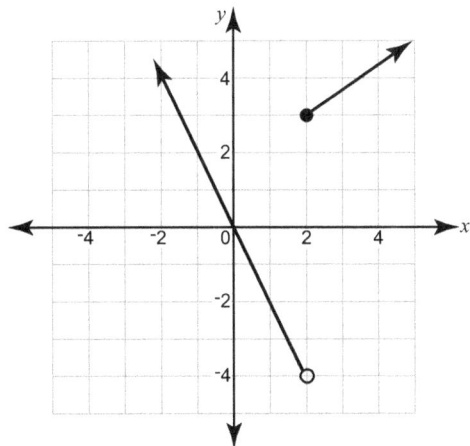

Inverse variation

If two parameters vary inversely, then, as one gets larger, the other one gets smaller. Instead, if x and y vary inversely, there is a constant c such that $xy = c$.

Example: the greater the speed at which you drive, the shorter the time it takes to get to Grandma's house, 120 miles away. This is an inverse variation. If the speed in miles per hour is plotted as x and the driving time in hours is plotted as y, then $xy=120$, as in the graph below.

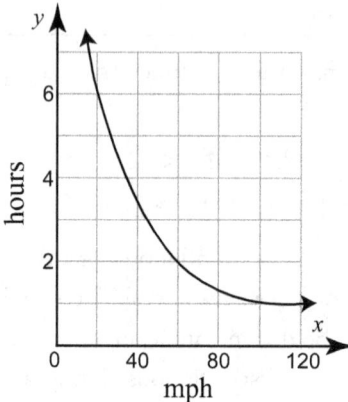

Example: If $30 were paid for 5 hours work, how much would be paid for 19 hours work? This is direct variation and $30 = 5c$, so the constant c is 6 ($6/hour). So $y = 6(19) = \$114$.

This could also be done as a proportion:
$$\frac{\$30}{5} = \frac{y}{19}$$
$$5y = \$570$$
$$y = \$114$$

Rational functions

A rational function $r(x)$ can be written as the ratio of two polynomial expressions $p(x)$ and $q(x)$ where $q(x)$ is nonzero.
$$r(x) = \frac{p(x)}{q(x)}, q(x) \neq 0$$
Examples of rational functions (and their associated expressions) are
$$r(x) = \frac{x^2 + 2x + 4}{x - 3} \text{ and } r(x) = \frac{x}{x^2 + 1}$$
Each of these examples is clearly the ratio of two polynomials. The following, however, is also a rational expression.
$$f(x) = \frac{1}{x + \frac{2}{x}}$$
This function can be shown to be a rational expression by converting it to standard form.

$$f(x) = \frac{1}{x + \frac{2}{x}}\left(\frac{x}{x}\right) = \frac{x}{x^2 + 2}$$

Since rational functions involve a denominator that is a polynomial expression (and not simply a constant), complicated division may be required to evaluate the function. Rational expressions are just like fractions and can be changed into other equivalent fractions through similar methods.

Graphing rational functions using asymptotes

A function may have one or more asymptotes. An asymptote is a line for which the distance between it and a function or curve is arbitrarily small, especially as the function tends toward infinity in some direction. Asymptotes can be either vertical, horizontal or slant. Consider, for instance, the plot of the hyperbola defined as follows.

$$g(x) = \pm\sqrt{x^2 + 1}$$

Note, for instance, that as x tends toward infinity, $g(x)$ gets arbitrarily close to x. The graph of $g(x)$ is shown below.

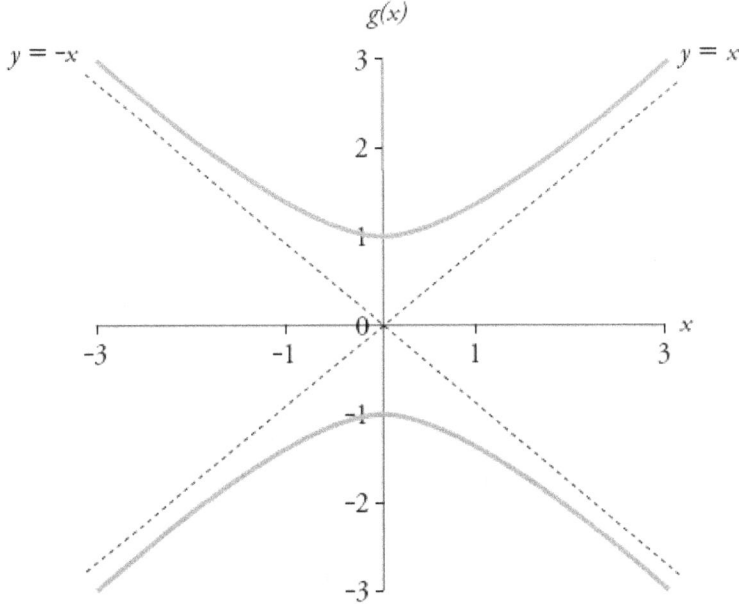

The (slant) asymptotes and their associated functions for this relation are displayed in the graph above as dashed lines.

Intervals of increase or decrease for a function are those regions over which the function is continuously increasing or decreasing, respectively. An interval of increase for a function $f(x)$ corresponds to any subset of the domain in which the slope of the function is always greater than zero; thus, if $x_2 > x_1$, then $f(x_2) > f(x_1)$ for any x_1 and x_2 in that interval of the function's domain. Likewise, an interval of decrease corresponds to any subset of the domain in which the slope of f is less than zero.

Operations on rational expressions

The rules for operating on rational expressions are the same as those for numerical fractions.

Addition and subtraction: To add and subtract rational expressions, find a lowest common denominator (LCD), rewrite all the expressions using the LCD, then add or subtract and simplify if possible.

Example: $\dfrac{1}{x+4} - \dfrac{3}{x-1}$

$LCD = (x+4)(x-1)$

$\dfrac{1}{x+4} - \dfrac{3}{x-1} = \dfrac{1(x-1)}{(x+4)(x-1)} - \dfrac{3(x+4)}{(x+4)(x-1)}$ Rewrite expressions using LCD.

$= \dfrac{x-1}{(x+4)(x-1)} - \dfrac{3x+12}{(x+4)(x-1)}$ Simplify.

$= \dfrac{x-1-(3x+12)}{(x+4)(x-1)}$ Subtract numerators over common denominator.

$= \dfrac{-2x-13}{(x+4)(x-1)}$ Simplify.

Multiplication: As with fractions, rational expressions can be multiplied simply by multiplying the numerators and multiplying the denominators.

Example: $\dfrac{2a}{a+b}\left(\dfrac{a-b}{c}\right) = \dfrac{2a(a-b)}{(a+b)c} = \dfrac{2a^2 - 2ab}{ac+bc}$

Division: The rule is the same as for fractions: invert the divisor (the second term) and multiply.

Example:

$$\frac{a^2-a}{b^3} \div \frac{a-1}{b} = \frac{a^2-a}{b^3} \times \frac{b}{a-1}$$ Invert divisor and multiply.

$$= \frac{a(a-1)b}{b^3(a-1)}$$ Multiply numerators, multiply denominators.

$$= \frac{a}{b^2}$$ Cancel common factors from numerator and denominator.

Simplifying rational expressions

To reduce a rational expression with more than one term in the denominator, the expression must be factored first. Factors that are the same will cancel. Addition or subtraction of rational expressions may first require finding a common denominator.

The first step to this end is to factor the denominators of both expressions to find the common factors. Then, proceed to rewrite the expressions with the common denominator by using the same methods as are used for numerical fractions.

Example: Rewrite the following fraction with a denominator of $(x+3)(x-5)(x+4)$:

$$\frac{x+2}{x^2+7x+12}$$

First, factor the denominator.

$$\frac{x+2}{x^2+7x+12} = \frac{x+2}{(x+3)(x+4)}$$

Multiply both the numerator and denominator by $(x-5)$: x 12

$$\frac{x+2}{x^2+7x+12} = \frac{x+2}{(x+3)(x+4)}\left(\frac{x-5}{x-5}\right) = \frac{(x+2)(x-5)}{(x+3)(x-5)(x+4)}$$

Although it is not necessary, the numerator and denominator can be multiplied out to represent the result as a rational expression in terms of polynomials.

$$\frac{x+2}{x^2+7x+12} = \frac{x^2-3x-10}{x^3+2x^2-23x-60}$$

The use of common denominators is helpful for addition and subtraction of rational expressions. Multiplication and division of rational expressions follows the standard rules of those operations.

Example: evaluate the following expression.

$$\frac{5}{x^2-9} - \frac{2}{x^2+4x+3}$$

Let the expression above be labeled $f(x)$ First, find the common denominator, then subtract appropriately.

$$f(x) = \frac{5}{(x+3)(x-3)} - \frac{2}{(x+3)(x+1)}$$

$$= \frac{5(x+1)}{(x+3)(x-3)(x+1)} - \frac{2(x-3)}{(x+3)(x-3)(x+1)}$$

$$= \frac{5(x+1)-2(x-3)}{(x+3)(x-3)(x+1)} = \frac{5x+5-2x+6}{(x+3)(x-3)(x+1)}$$

$$= \frac{3x+11}{(x+3)(x-3)(x+1)} = \frac{3x+11}{x^3+x^2-9x-9}$$

The above expression is the result, both in factored form and in standard form.

Example: evaluate the following expression.

$$\frac{x^2-2x-24}{x^2+6x+8}\left(\frac{x^2+3x+2}{x^2-13x+42}\right)$$

Label the expression as $f(x)$ First, factor each polynomial, simplifying as appropriate, then multiply.

$$f(x) = \frac{(x-6)(x+4)}{(x+2)(x+4)} \cdot \frac{(x+1)(x+2)}{(x-6)(x-7)}$$

$$= \frac{x+1}{x-7}$$

Solving equations involving rational expressions

To solve an equation containing rational expressions, set the expression equal to zero (which leads to the elimination of the denominator) and solve, as with simple polynomials.

$$r(x) = \frac{p(x)}{q(x)}$$

Note, however, that solutions to $p(x) = 0$ may lead to undefined values for $r(x)$, (that is, values for which $q(x) = 0$), and must be checked prior to acceptance.

This difficulty can be alleviated to some extent by factoring $p(x)$ and $q(x)$ and eliminating common factors.

Example: Find the solutions for $\dfrac{12}{2x^2 - 4x} + \dfrac{13}{5} = \dfrac{9}{x - 2}$

Factor and rearrange the equation as follows, then solve for x.

$$\frac{12}{2x(x-2)} - \frac{9}{x-2} = -\frac{13}{5}$$

$$\frac{12}{2x(x-2)} - \frac{9(2x)}{2x(x-2)} = \frac{-18x + 12}{2x(x-2)} = -\frac{13}{5}$$

$$-18x + 12 = -\frac{13}{5}(2x)(x-2) = -\frac{26}{5}x^2 + \frac{52}{5}x$$

$$\frac{26}{5}x^2 - \frac{52}{5}x - 18x + 12 = 0$$

$$0 = \frac{26}{5}x^2 - \frac{142}{5}x + 12 = 26x^2 - 142x + 60 = 13x^2 - 71x + 30$$

The solutions for x can be found by factoring the above expression.
$$13x^2 - 71x + 30 = (x - 5)(13x - 6) = 0$$

Thus, $x = 5$ or $x = \dfrac{6}{13}$. These solutions can be confirmed by substitution into the original equation.

Polynomial functions

A polynomial is a sum of terms, where each term is a constant multiplied by a variable raised to a positive integer power. The general form of a polynomial $P(x)$ is:

$$a_n x^n + a_{n-1} x^{n-1} + \ldots + a_2 x^2 + a_1 x + a_0$$

Polynomials written in standard form have the terms written in order of decreasing exponent value, as shown above. The degree of a polynomial function in one variable is the value of the largest exponent to which the variable is raised. The above expression is a polynomial of degree n (assuming that $a_n \neq 0$). Any function that represents a line, for instance, is a polynomial function of degree 1. Quadratic functions are polynomials of degree 2.

There are many methods for solving problems that involve polynomial equations.

For instance, in cases in which a polynomial is highly complicated or involves constants that do not permit methods such as factoring, a numerical approach may be appropriate. Newton's method is one possible approach to solving a polynomial equation numerically. At other times, solving a polynomial equation may require a graphical approach whereby the behavior of the function is examined on a visual plot. When using Newton's method, graphing the function can be helpful for estimating the locations of the roots (if any).

Polynomial equations with real coefficients cannot always be solved using only real numbers.

Example: consider the quadratic function given below:

$$f(x) = x^2 + 1$$

There are no real roots for this equation, since

$$x^2 + 1 = 0 \rightarrow x^2 = -1.$$

The Fundamental Theorem of Algebra (see below), however, indicates that there must be two (possibly non-distinct) solutions to this equation. Note that if the complex numbers are permitted as solutions to this equation, then $x = \pm i$.

Thus, generally, solutions to any polynomial equation with real coefficients exist in the set of complex numbers.

If a phenomenon or situation can be modeled with a polynomial equation, the following theorems can be helpful in solving the equation.

These theorems include the Fundamental Theorem of Algebra, the Factor Theorem, the Complex Conjugate Root Theorem, and the Rational Root Theorem.

The fundamental theorem of algebra

The fundamental theorem of algebra states that a polynomial expression of degree n must have n roots (which may be real or complex and which may not be distinct). It follows from the theorem that if the degree of a polynomial is odd, then it must have at least one real root.

Polynomial functions are in the form of $P(x)$ given below, where n is the degree of the polynomial and the constant *an* is nonzero.

$$P(x) = a_n x^n + a_{n-1} x^{n-1} + ... + a_2 x^2 + a_1 x + a_0$$

If $P(c) = 0$ for some number c, then c is said to be a *zero* (or *root*) of the function.

A zero is also called a *solution* to the equation $P(x) = 0$.

The existence of n solutions can be seen by looking at a factorization of $P(x)$.

Example: find the roots of $P(x) = x^2 - x - 6$.
This second-degree polynomial can be factored into
$P(x) = (x+2)(x-3)$.
Note that $P(x)$ has two roots in this case: $x = -2$ and $x = 3$. This corresponds to the degree of the polynomial, $n = 2$. In some cases, however, there may be nondistinct roots. Consider
$P(x) = x^2$.
$P(x) = x(x)$
Note that the polynomial is factored in the same way as the previous example, but, in this case, the roots are identical: $x = 0$. Thus, although there are two roots for this second-degree polynomial, the roots are not distinct.
Likewise, roots of a polynomial may be complex.

Example: find the roots of $P(x) = x^2 + 1$.
The range of this function is $P(x) \geq 1$, so there are no real roots, since the function never crosses the x-axis. Nevertheless, if complex values of x are permitted, there are cases where $P(x)$ is zero. Factor $P(x)$ as before, but this time use complex numbers.

The solutions are $x = i$ and $x = -i$. Thus, this second-degree polynomial still has two roots.

The factor theorem

The factor theorem establishes the relationship between the factors and the zeros or roots of a polynomial and is useful for finding the factors of higher-degree polynomials. The theorem states that a polynomial $P(x)$ $(x - c)$ if and only if $P(c) = 0$.

For a general nth-degree polynomial, the function $P(x)$ can therefore be factored as follows:

$$P(x) = (x - c_n)(x - c_{n-1})...(x - c_2)(x - c_1)$$

As with the second-degree polynomial examples examined above, a general nth-degree polynomial can have roots c_i that are distinct or nondistinct, and real or complex. Since this is the case, if all of the roots of a polynomial are known, then a function $P(x)$ is determined based on the factoring approach shown above.

In addition, if a single root c is known, then the polynomial can be simplified (that is, it can be reduced by one degree) using division.

$$Q(x) = \frac{P(x)}{x - c}$$

Here, if $P(x)$ has degree n, then $Q(x)$ has degree $n - 1$. If some number of roots are known, the task of finding the remainder of the roots can be simplified by performing the division represented above. As each successive root is found, the degree of the polynomial can be reduced to further simplify finding the remainder of the roots.

Factorials

The factorial of a positive integer is the product of that integer and every lesser positive integer down to 1. The factorial of 5, for instance, which is written as 5!, equals $5 \times 4 \times 3 \times 2 \times 1$.

Permutations and combinations

A permutation is one of the possible arrangements of n items, without repetition, where the order of selection is important.

A combination is one of the possible arrangements of n items, without repetition, where the order of selection is not important.

Example: If any two numbers are selected from the set {1, 2, 3, 4}, list the possible permutations and combinations.

Combinations   Permutations

12, 13, 14, 23, 24, 34   12, 21, 13, 31, 14, 41, 23, 32, 24, 42, 34, 43

six ways   twelve ways

Note that the list of permutations includes 12 and 21 as separate possibilities since the order of selection is important. In the case of combinations, however, the order of selection is not important and, therefore, 12 is the same combination as 21. Hence, 21 is not listed separately as a possibility.

The number of permutations and combinations may also be found by using the formulae given below.

The number of possible permutations in selecting r objects from a set of n objects is given by

$$_nP_r = \frac{n!}{(n-r)!}$$

The notation $_nP_r$ is read "the number of permutations of n objects taken r at a time."

In our example, two objects are being selected from a set of four.

$$_4P_2 = \frac{4!}{(4-2)!} = \frac{4 \times 3 \times 2 \times 1}{2 \times 1} = \frac{24}{2} = 12$$

The number of possible combinations in selecting r objects from a set of n objects is given by

$$_nC_r = \frac{n!}{(n-r)!r!}$$

In our example,

$$_4C_2 = \frac{4!}{(4-2)!2!} = \frac{24}{2(2)} = 6$$

Permutations and combinations

Objects arranged in a row

It can be shown that $_nP_n$, the number of ways n objects can be arranged in a row, is equal to $n!$ We can imagine n positions being filled, one at a time. The first position can be filled in n ways using any one of the n objects.

Since one of the objects has been used, the second position can be filled in only $(n-1)$ ways. Similarly, the third position can be filled in $(n-2)$ ways, and so on. Hence, the total number of possible arrangements of n objects in a row is given by

$$_nP_n = n(n-1)(n-2)\ldots 1 = n!$$

Example: Five books are placed in a row on a bookshelf. In how many different ways can they be arranged?

The number of possible ways in which 5 books can be arranged in a row is $5! = 5 \times 4 \times 3 \times 2 \times 1 = 120$

The formula given above for $_nP_r$, the number of possible permutations of r objects selected from n objects, can also be proved in a similar manner. If r positions are filled by selecting from n objects, the first position can be filled in n ways, the second position can be filled in $n-1$ ways, and so on (as shown before). The rth position can be filled in $n-(r-1) = n-r+1$ ways. Hence,

$$_nP_r = n(n-1)(n-2)\ldots(n-r+1) = \frac{n!}{(n-r)!}$$

The formula for the number of possible combinations of r objects selected from n objects, $_nC_r$, may be derived by using the above two formulae. For the same set of r objects, the number of permutations is $r!$. All these permutations, however, correspond to the same combination. Hence,

$$_nC_r = \frac{_nP_r}{r!} = \frac{n!}{(n-r)!r!}$$

CLEP College Algebra
Sample Test 1

Sample Test One

Direction: Read each item and select the best response.

1. Which of the following is a factor of the expression $9x^2 + 6x - 35$?

    [A] $3x - 5$

    [B] $3x - 7$

    [C] $x + 3$

    [D] $x - 2$

    [E] $x - 3$

2. Given $f(x) = 3x - 2$ and $g(x) = x^2$, determine $g(f(x))$.

    [A] $3x^2 - 2$

    [B] $9x^2 + 4$

    [C] $9x^2 - 12x + 4$

    [D] $3x^3 - 2$

    [E] $9x^2 - 36$

3. Solve for x: $18 = 4 + |2x|$

    [A] $\{-11, 7\}$

    [B] $\{-7, 0, 7\}$

    [C] $\{-7, 7\}$

    [D] $\{-11, 11\}$

    [E] $\{-8, 8\}$

4. Solve for x by factoring: $2x^2 - 3x - 2 = 0$

   [A] $x = (-1, 2)$

   [B] $x = (0.5, -2)$

   [C] $x = (-0.5, 2)$

   [D] $x = (1, -2)$

   [E] $x = (-2, 2)$

5. Which of the following illustrates an inverse property?

   [A] $a + b = a - b$

   [B] $a + b = b + a$

   [C] $a + 0 = a$

   [D] $a + (-a) = 0$

   [E] $b - a = 0$

6. The conjugate of $4 + 5i$ is

   [A] $-4 + 5i$

   [B] $4 - 5i$

   [C] $4i + 5$

   [D] $4i - 5$

   [E] $-4 - 5i$

7. Simplify: $(6+3i)-(4-2i)$

 [A] $2+5i$

 [B] $2+i$

 [C] $10+5i$

 [D] $2-2i$

 [E] $10-5i$

8. Simplify: $\dfrac{10}{1+3i}$

 [A] $-1.25(1-3i)$

 [B] $1.25(1+3i)$

 [C] $1+3i$

 [D] $1-3i$

 [E] $10+3i$

9. Solve $(2b^3 \cdot b^2)^3$

 [A] $3b^9$

 [B] $2b^8$

 [C] $8b^{15}$

 [D] $2b^{18}$

 [E] $8b^{18}$

10. Which of the following is incorrect?

[A] $\left(x^2 y^3\right)^2 = x^4 y^6$

[B] $m^2(2n)^3 = 8m^2 n^3$

[C] $\dfrac{m^3 n^4}{m^2 n^2} = mn^2$

[D] $\left(x + y^2\right)^2 = x^2 + y^4$

[E] $\left(2s^{-4} w^4\right)\left(7sw^{-5}\right) = \dfrac{14}{s^3 w}$

11. Evaluate $3^{\frac{1}{2}}\left(9^{\frac{1}{3}}\right)$

[A] $27^{\frac{5}{6}}$

[B] $9^{\frac{7}{12}}$

[C] $3^{\frac{5}{6}}$

[D] $3^{\frac{6}{7}}$

[E] $9^{\frac{12}{7}}$

12. Simplify: $\dfrac{4x^0 y^{-2} z^3}{4x}$

[A] $\dfrac{z^3}{y^2}$

[B] $\dfrac{z^3}{y^2 x}$

[C] $\dfrac{z^2}{y^3}$

[D] $\dfrac{z^3}{x^2 y}$

[E] $z^3 y^2$

13. The exponential equation $2^5 = 32$ can be written as:

    [A] $\log_2(5) = 32$

    [B] $\log_{10}(32) = 5$

    [C] $\log_5(32) = 2$

    [D] $\log_2(32) = 5$

    [E] $\log_5(2) = 32$

14. Which equation corresponds to the logarithmic statement $\log_x k = m$?

    [A] $x^m = k$

    [B] $k^m = x$

    [C] $x^k = m$

    [D] $m^x = k$

    [E] $k^x = m$

15. Solve for x: $\log_6(x-5) + \log_6 x = 2$

    [A] $x = 9$

    [B] $x = 2, x = 7$

    [C] $x = 6$

    [D] $x = -2, x = -7$

    [E] $x = -4, x = -9$

16. Solve for the slope m and y-intercept: $3x+2y=14$

   [A] $m=\frac{2}{3}, y=5$

   [B] $m=-\frac{3}{2}, y=7$

   [C] $m=\frac{3}{2}, y=-7$

   [D] $m=-\frac{2}{3}, y=-5$

   [E] $m=2, y=7$

17. Simplify: $-4(-4x-1)-4(7x+3)$

   [A] $-44x+16$

   [B] $12x-16$

   [C] $44x-16$

   [D] $-12x-8$

   [E] $-11x+2$

18. Solve $-2x<5$.

   [A] $x<-\frac{5}{2}$

   [B] $x>-\frac{2}{5}$

   [C] $x>-\frac{5}{2}$

   [D] $x>\frac{5}{2}$

   [E] $x<\frac{5}{2}$

19. Solve $10 \leq 3x + 4 \leq 19$.

   [A] $2 \leq x \leq 5$

   [B] $-2 \leq x \leq 5$

   [C] $x \leq 5$

   [D] $x \geq 2$

   [E] $-5 \leq x \leq -2$

20. Solve for x: $x^2 + 10x - 24 = 0$

   [A] $(-5, 12)$

   [B] $(-10, 8)$

   [C] $(12, 2)$

   [D] $(10, 8)$

   [E] $(-12, 2)$

21. Find a quadratic equation with roots of 4 and -9.

   [A] $x^2 - 5x + 36 = 0$

   [B] $x^2 + 5x - 36 = 0$

   [C] $4x^2 - 9x - 5 = 0$

   [D] $x^2 + 4x - 9 = 0$

   [E] $5x^2 - 9x + 4 = 0$

22. **Solve:** $4800 \leq 200x - 2x^2$

   [A] $-40 \leq x \leq 40$

   [B] $x \leq 40$

   [C] $40 \leq x \leq 60$

   [D] $x = 40$

   [E] $x = -40$

23. **Solve:** $|3x + 2| = 4x + 5$

   [A] $x = -3$

   [B] $x = -1$

   [C] $x = 3$

   [D] $x = 1$

   [E] $x = 6$

24. **Solve:** $|3x - 5| = \dfrac{1}{2}$

   [A] $x = -\dfrac{11}{6}, -\dfrac{3}{2}$

   [B] $x = -\dfrac{11}{6}, \dfrac{3}{2}$

   [C] $x = \dfrac{11}{6}, -\dfrac{3}{2}$

   [D] $x = \dfrac{11}{6}, \dfrac{3}{2}$

   [E] $x = 11, \dfrac{3}{2}$

25. Solve: $2|3x+9|<36$

   [A] $x<-9$

   [B] $x>3$

   [C] $3<x<9$

   [D] $-9<x<-3$

   [E] $-9<x<3$

26. Solve for x and y:
   $4x+3y=-1$
   $5x+4y=1$

   [A] $x=-7, y=9$

   [B] $x=7, y=-9$

   [C] $x=7, y=9$

   [D] $x=-7, y=-9$

   [E] $x=y=7$

27. Which point is in the solution set for the system of inequalities below?
   $x-7>1$
   $y<2x-1$

   [A] $(-1,-1)$

   [B] $(-2,-1)$

   [C] $(0,1)$

   [D] $(0,-2)$

   [E] $(1,1)$

28. Solve: $3^{2x-1} = 27$

   [A] $x = 2$

   [B] $x = -3$

   [C] $x = -2$

   D. $x = 3$

   [E] $x = \dfrac{2}{3}$

29. Solve: $\log_b(x^2) = \log_b(2x - 1)$

   [A] $x = -2$

   [B] $x = 1$

   [C] $x = -1$

   [D] $x = 2$

   [E] $x = 4$

30. Solve: $\log_2(x) + \log_2(x - 2) = 3$

   [A] $x = 4$

   [B] $x = -4, 2$

   [C] $x = -4, -2$

   [D] $x = 4, -2$

   [E] $x = 2$

31. If $f(x) = -3x + 8$, find $f(5)$.

   [A] 23

   [B] -23

   [C] 7

   [D] -7

   [E] 21

32. Find the zeros of the function $h(x) = \dfrac{x-9}{x+2}$.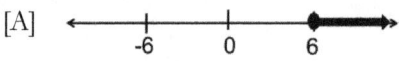

   [A] $\{9\}$

   [B] $\{-2\}$

   [C] $\left\{-\dfrac{9}{2}\right\}$

   [D] $\{-2, 9\}$

   [E] This function has no zeros.

33. Which number line shows the solution to $7x - 5 \geq 9x - 17$?

   [A]

   [B]

   [C]

   [D]

   [E]

34. Which graph represents the equation of $y = x^2 + 3x$?

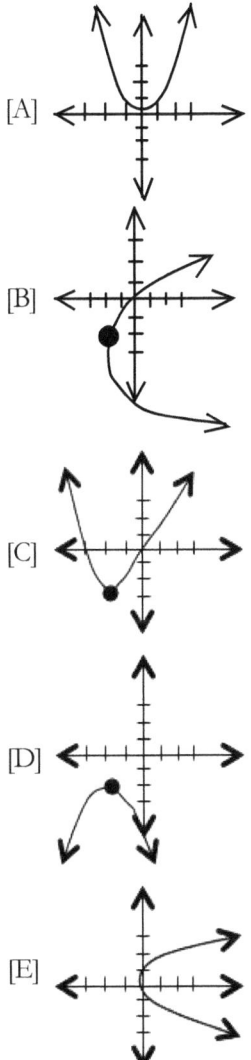

35. Based on the given table, if $y_1 = x^3$, what is the equation for y_2?

x	-2	-1	0	1	2	3
y_1	-8	-1	0	1	8	27
y_2	-18	-11	-10	-9	-2	-17

[A] $y_2 = x^5$

[B] $y_2 = -x^3$

[C] $y_2 = (-x)^3$

[D] $y_2 = (x-10)^3$

[E] $y_2 = x^3 - 10$

36. **Identify the domain and range of the relation:**
{(2,−5),(4,31),(11,−11),(−21,3)}

[A] Domain is {−21}, range is {−11}.

[B] Domain is {−5,31,−11,3}, range is {2,4,11,−21}.

[C] Domain is {11}, and range is {31}.

[D] Domain and range are indeterminate.

[E] Domain is {2,4,11,−21}, range is {−5,31,−11,3}.

37. Determine the domain of $y = -\sqrt{-2x+3}$.

[A] $x = 3$

[B] $x \leq \dfrac{3}{2}$

[C] $x > \dfrac{3}{2}$

[D] $x = 2$

[E] $x = 0$

38. For the function $h(x)$ whose graph is shown below, select the domain and range.

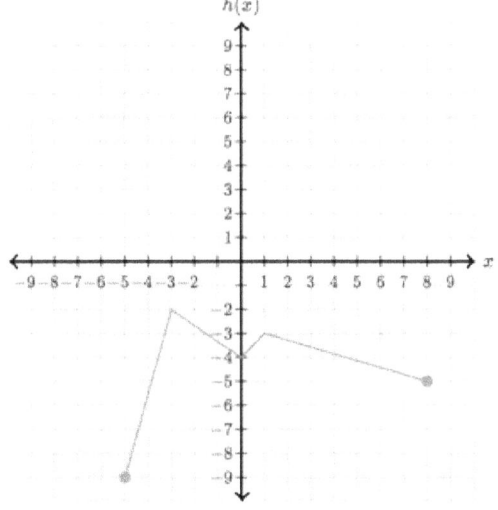

[A] Domain is $-5 \leq x \leq 8$, range is $-9 \leq h \leq -2$.

[B] Domain is -5, range is -5.

[C] Range is $-5 \leq x \leq 8$, domain is $-9 \leq h \leq 2$.

[D] Domain is $x \geq -5$, range is $h \geq -9$.

[E] Domain is 8, range is -2.

39. Given $f(x) = 3x^2 - 7x + 5$, find $f(-4)$.

[A] −71

[B] 25

[C] 81

[D] −25

[E] 71

40. For $h(x) = 3x^2 + ax - 1$, $h(3) = 8$, find the value of a.

[A] 6

[B] −6

[C] −18

[D] 18

[E] 27

41. Given $f(x) = 3x^2 - 7x + 5$, find $\dfrac{f(x+h) - f(x)}{h}$

[A] $7h$

[B] $6xh - 7$

[C] $6x + 3h - 7$

[D] $3x + 6h + 7$

[E] $5x$

42. Find the *x*- and *y*- intercepts for $5x - 3y = 15$.

   [A] $x = 0, y = 0$

   [B] $x = -3, y = 5$

   [C] $x = -1, y = 5$

   [D] $x = -5, y = 3$

   [E] $x = 3, y = -5$

43. Which of the figures is a reflection of the triangle shown?

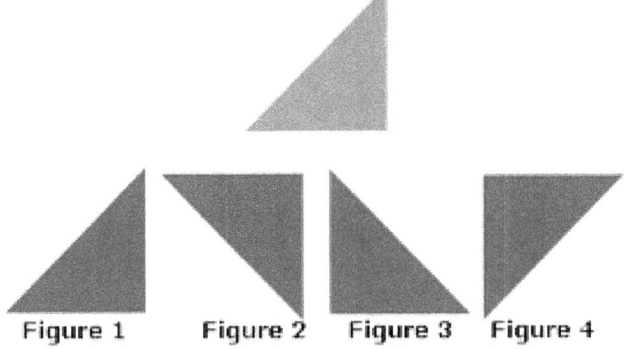

   [A] Figure 1 and Figure 4

   [B] Figure 4 and Figure 3

   [C] Figure 2 and Figure 1

   [D] Figures 2, 3 and 4

   [E] Figure 1 and Figure 2

44. Name the transformation shown.

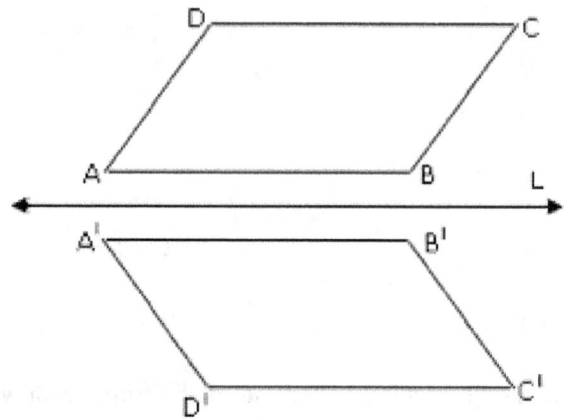

[A] Translation

[B] Rotation

[C] Reflection

[D] Dilation

[E] Cannot be determined

45. Find the inverse of $y = 3x - 2$.

[A] $y = \dfrac{1}{3x-2}$

[B] $y = \dfrac{x+2}{3}$

[C] $x = \dfrac{y+2}{3}$

[D] $y - 3x - 2 = 0$

[E] $3x - y - 2 = 0$

46. Find the inverse of $f(x) = -\frac{1}{3}x + 1$

    [A] $f^{-1}(x) = 1$

    [B] $f^{-1}(x) = 3x$

    [C] $f^{-1}(x) = 3x - 3$

    [D] $f^{-1}(x) = -3x + 3$

    [E] $f^{-1}(x) = x^2$

47. If $f(x) = 3x - 2$ and $g(x) = \frac{x}{3} + \frac{2}{3}$, which of the following is true:

    [A] $f(x)$ is the inverse of $g(x)$.

    [B] $f(x) = g^{-1}(x)$

    [C] There is no connection between $f(x)$ and $g(x)$.

    [D] $g(x) = f(x)$

    [E] A and B

48. Identify the real numbers in the list: $1.67, \pi, \sqrt{5}, 0$

    [A] All

    [B] $1.67, \sqrt{5}, 0$

    [C] $1.67, 0$

    [D] 0

    [E] None

49. Which of the following is false?

   [A] Every rational number is a real number.

   [B] Every imaginary number is a real number.

   [C] Every integer is a whole number.

   [D] Every integer is a real number.

   [E] Every natural number is positive.

50. Which selection below is NOT a real number?

   [A] −3

   [B] 0.6666...

   [C] $\dfrac{\pi}{2}$

   [D] $3+\sqrt{2}$

   [E] 3i

51. Simplify $\sqrt{-9}$.

   [A] -3

   [B] -3i

   [C] 3i

   [D] 3

   [E] 0

52. Simplify $(i)(2i)(-3i)$.

 [A] $6i$

 [B] $-6i^3$

 [C] $-6i$

 [D] 0

 [E] $6i^3$

53. Simplify i^{17}.

 [A] $17i$

 [B] i

 [C] $-17i$

 [D] $-i$

 [E] 1

54. List the first four terms of the following sequence, beginning with $n=0$.
$$A_n = \frac{(-1)^n}{(n+1)!}$$

 [A] $\frac{1}{2}, 1, \frac{3}{2}, 2$

 [B] $-1, -\frac{1}{2}, 0, \frac{1}{2}$

 [C] $0, 1, 2, 3$

 [D] $1, -\frac{1}{2}, \frac{1}{6}, -\frac{1}{24}$

 [E] $0, -1, -\frac{1}{2}, \frac{2}{3}$

55. Expand the following series and find the sum:

$$\sum_{n=0}^{4} 2n$$

[A] 20

[B] 8

[C] 16

[D] 4

[E] 32

56. Write the series in sigma notation: $-3+0+9+24+45+72+105$

[A] $\sum_{a=0}^{6} 3a^2$

[B] $\sum_{a=0}^{6} 3a^2 - 3$

[C] $\sum_{a=0}^{6} a^2 - 3$

[D]. $\sum_{a=1}^{6} 3a^2 - 1$

[E] $\sum_{a=0}^{5} a^2 - 3$

57. Find $\dfrac{8!}{6!2!}$

[A] $\dfrac{2}{3}$

[B] $\dfrac{4}{6}$

[C] 28

[D] 48

[E] 24

58. Expand the binomial $(2x+3y)^4$

[A] $16x^4 + 24x^3y + 36x^2y^2 + 54xy^3 + 81y^4$
    $2x^4 + 6x^3y + 6x^2y^2 + 6xy^3 + 3y^4$

[B] $16x^4 + 81y^4$

[C] 16x4 + 96x3y + 216x2y2 + 216xy3 + 81y4

[D] $16x^4 + 24x^3y^3 + 36x^2y^2 + 54xy + 81y^4$

[E] $x^4 + 4x^3y + 6x^2y^2 + 4xy^3 + y^4$

59. Evaluate the determinant of the matrix:
$$\begin{pmatrix} -2 & 4 \\ -4 & 3 \end{pmatrix}$$

[A] 10

[B] -24

[C] 4

[D] -10

[E] 24

60. Evaluate the determinant of the matrix for $y = 4$. $\begin{pmatrix} -5y & 3y \\ y-1 & y-3 \end{pmatrix}$

[A] 35

[B] 12

[C] -56

[D] -12

[E] 56

SAMPLE TEST 1 ANSWER KEY

Question Number	Correct Answer	Your Answer
1	A	
2	C	
3	C	
4	C	
5	D	
6	B	
7	A	
8	D	
9	C	
10	D	
11	C	
12	B	
13	D	
14	A	
15	A	
16	B	
17	D	
18	C	
19	A	
20	E	
21	B	
22	C	
23	B	
24	D	
25	E	
26	A	
27	D	
28	A	
29	B	
30	A	

Question Number	Correct Answer	Your Answer
31	D	
32	A	
33	B	
34	C	
35	E	
36	E	
37	B	
38	A	
39	C	
40	B	
41	C	
42	E	
43	D	
44	C	
45	B	
46	D	
47	E	
48	A	
49	E	
50	E	
51	C	
52	A	
53	B	
54	D	
55	A	
56	B	
57	C	
58	C	
59	A	
60	C	

RATIONALES

1. Which of the following is a factor of the expression $9x^2 + 6x - 35$?

   [A] $3x - 5$

   [B] $3x - 7$

   [C] $x + 3$

   [D] $x - 2$

   [E] $x - 3$

   The answer is: A

   The trinomial can be factored into two binomials, one with addition and one containing subtraction. The factors of 9 to use are 3 and 3 and 7 and 5 are used for 35.

   $(3x - 5)(3x + 7)$ checks when multiplying back through: 9x2 + 21x − 15x − 35 = $9x^2 + 6x - 35$

2. Given $f(x) = 3x - 2$ and $g(x) = x^2$, determine $g(f(x))$.

   [A] $3x^2 - 2$

   [B] $9x^2 + 4$

   [C] $9x^2 - 12x + 4$

   [D] $3x^3 - 2$

   [E] $9x^2 - 36$

   The answer is: C

   Evaluate: $g(f(x)) = g(3x - 2) = (3x - 2)^2$

   Simplify by expanding: (3x-2)(3x-2)

   $9x^2 - 6x - 6x + 4$ which simplifies to choice C

3. Solve for x: $18 = 4 + |2x|$

   [A] $\{-11, 7\}$

   [B] $\{-7, 0, 7\}$

   [C] $\{-7, 7\}$

   [D] $\{-11, 11\}$

   [E] $\{-8, 8\}$

   The answer is: C
   First isolate the absolute value: $18 = 4 + |2x|$
   $$14 = |2x|$$
   Then use the definition of absolute value to set up and solve two equations: $2x = 14$ or $2x = -14$
   $$x = 7, \quad x = -7$$

4. Solve for x by factoring: $2x^2 - 3x - 2 = 0$

   [A] $x = (-1, 2)$

   [B] $x = (0.5, -2)$

   [C] $x = (-0.5, 2)$

   [D] $x = (1, -2)$

   [E] $x = (-2, 2)$

   The answer is: C

   Factor the trinomial into one binomial sum and one binomial difference:
   $(2x+1)(x-2)$

   Then set each factor equal to zero and solve for x:
   $2x + 1 = 0$ or $x - 2 = 0$
   $$x = -\frac{1}{2}, 2$$

   114   CLEP Algebra

5. Which of the following illustrates an inverse property?

   [A] $a+b=a-b$

   [B] $a+b=b+a$

   [C] $a+0=a$

   [D] $a+(-a)=0$

   [E] $b-a=0$

   The answer is: D

   Choice D represents the sum of a number and its opposite, or additive inverse. This illustrates the inverse property.

6. The conjugate of $4+5i$ is

   [A] $-4+5i$

   [B] $4-5i$

   [C] $4i+5$

   [D] $4i-5$

   [E] $-4-5i$

   The answer is: B

   For any complex number $a + bi$, the conjugate is defined as $a - bi$.

7. Simplify: $(6+3i)-(4-2i)$

   [A] $2+5i$

   [B] $2+i$

   [C] $10+5i$

   [D] $2-2i$

   [E] $10-5i$

   The answer is: A

   To add complex numbers, add the real parts together and the imaginary parts together. (6+3i)+-(4-2i)=6+(-4)+3i+2i=2+5i

8. Simplify: $\dfrac{10}{1+3i}$

   [A] $-1.25(1-3i)$

   [B] $1.25(1+3i)$

   [C] $1+3i$

   [D] $1-3i$

   [E] $10+3i$

   The answer is: D

   A rational expression with an imaginary denominator must be simplified using the conjugate of the complex denominator:
   $$\frac{10}{1+3i} \cdot \frac{1-3i}{1-3i} = \frac{10-30i}{1-9i^2} = \frac{10-30i}{1+9} = \frac{10-30i}{10} = 1-3i$$

116   CLEP Algebra

9. Solve $(2b^3 \cdot b^2)^3$

   [A] $3b^9$

   [B] $2b^8$

   [C] $8b^{15}$

   [D] $2b^{18}$

   [E] $8b^{18}$

   The answer is: C

   First simplify inside the parenthesis by adding exponents:
   $(2b^3 \cdot b^2)^3 = (2b^5)^3$
   Then raise to the third power, multiplying exponents:
   $(2b^5)^3 = 2^3 b^{15} = 8b^{15}$

10. Which of the following is incorrect?

    [A] $(x^2 y^3)^2 = x^4 y^6$

    [B] $m^2(2n)^3 = 8m^2 n^3$

    [C] $\dfrac{m^3 n^4}{m^2 n^2} = mn^2$

    [D] $(x+y^2)^2 = x^2 + y^4$

    [E] $(2s^{-4}w^4)(7sw^{-5}) = \dfrac{14}{s^3 w}$

    The answer is: D

    A power can distribute to a monomial, as seen in choices A and E, but not to a binomial. To find the correct The answer is to D, expand and multiply: $(x+y^2)^2 = (x+y^2)(x+y^2) = x^2 + xy^2 + xy^2 + y^4$

11. Evaluate $3^{\frac{1}{2}}\left(9^{\frac{1}{3}}\right)$

   [A] $27^{\frac{5}{6}}$

   [B] $9^{\frac{7}{12}}$

   [C] $3^{\frac{5}{6}}$

   [D] $3^{\frac{6}{7}}$

   [E] $9^{\frac{12}{7}}$

   The answer is: C
   Rewrite the expression with like bases. $3^{\frac{1}{2}}\left(9^{\frac{1}{3}}\right) = 3^{\frac{1}{2}}\left(3^{2}\right)^{\frac{1}{3}}$

   Then use exponent rules to combine the like bases.
   $$3^{\frac{1}{2}}\left(3^{2}\right)^{\frac{1}{3}} = 3^{\frac{1}{2}}\left(3^{\frac{2}{3}}\right) = 3^{\left(\frac{3}{6}+\frac{4}{6}\right)} = 3^{\frac{5}{6}}$$

12. Simplify: $\dfrac{4x^0 y^{-2} z^3}{4x}$

   [A] $\dfrac{z^3}{y^2}$

   [B] $\dfrac{z^3}{y^2 x}$

   [C] $\dfrac{z^2}{y^3}$

   [D] $\dfrac{z^3}{x^2 y}$

   [E] $z^3 y^2$

   The answer is: B

   Initially, 4/4 reduces to 1 and x^0 also equals 1. Then the expression is
   $$\dfrac{y^{-2} z^3}{x} = \dfrac{z^3}{xy^2}$$

118   CLEP Algebra

13. The exponential equation $2^5 = 32$ can be written as:

[A] $\log_2(5) = 32$

[B] $\log_{10}(32) = 5$

[C] $\log_5(32) = 2$

[D] $\log_2(32) = 5$

[E] $\log_5(2) = 32$

The answer is: D

Logarithmic and exponential equations share the following relationship:
If $(base)^{exponent} = n$, then $log_{(base)} n = exponent$.

14. Which equation corresponds to the logarithmic statement $\log_x k = m$?

[A] $x^m = k$

[B] $k^m = x$

[C] $x^k = m$

[D] $m^x = k$

[E] $k^x = m$

The answer is: A

See explanation for question 13.

15. Solve for x: $\log_6(x-5) + \log_6 x = 2$

[A] $x = 9$

[B] $x = 2, x = 7$

[C] $x = 6$

[D] $x = -2, x = -7$

[E] $x = -4, x = -9$

The answer is: A

Use the log rule: $\log_b(a) + \log_b(c) = \log_b(ac)$ to simplify the equation.
$$\log_6(x-5) + \log_6 x = \log_6 x(x-5) = 2$$
Then rewrite the log as an exponential relationship and solve for x.
$$\log_6 x(x-5) = 2$$
$$6^2 = x(x-5)$$
$$36 = x^2 - 5x$$
$$0 = x^2 - 5x - 36$$
$$0 = (x-9)(x+4), x = 9 \text{ or } x = -4$$

However, this solution can yield extraneous solutions, so the answer is must be checked.

$\log_6(9-5) + \log_6 9 \stackrel{?}{=} 2$

$\log_6(4)(9) \stackrel{?}{=} 2$ $\qquad\qquad \log_6(-4-5) + \log_6(-4) \stackrel{?}{=} 2$

$\log_6 36 = 2$

The second portion of the check fails, as the log of a negative number is undefined. So the only solution to the problem is x = 9

16. **Solve for the slope** m **and** y**-intercept:** $3x+2y=14$

   [A] $m=\frac{2}{3}, y=5$

   [B] $m=-\frac{3}{2}, y=7$

   [C] $m=\frac{3}{2}, y=-7$

   [D] $m=-\frac{2}{3}, y=-5$

   [E] $m=2, y=7$

   The answer is: B

   Put the given equation into slope intercept form, y=mx + b, where m is the slope and b the y intercept.
   $$3x+2y=14$$
   $$2y=-3x+14$$
   $$y=-\frac{3}{2}x+7$$

17. **Simplify:** $-4(-4x-1)-4(7x+3)$

   [A] $-44x+16$

   [B] $12x-16$

   [C] $44x-16$

   [D] $-12x-8$

   [E] $-11x+2$

   The answer is: D

   Use the distributive property to begin simplifying; then collect like terms.
   $$-4(-4x-1)-4(7x+3)$$
   $$16x+4-28x-12$$
   $$-12x-8$$

18. Solve $-2x < 5$.

[A] $x < -\dfrac{5}{2}$

[B] $x > -\dfrac{2}{5}$

[C] $x > -\dfrac{5}{2}$

[D] $x > \dfrac{5}{2}$

[E] $x < \dfrac{5}{2}$

The answer is: C

To solve the inequality, divide both sides by -2. This step, however, requires a reversal of the inequality symbol, resulting in choice C.

19. Solve $10 \leq 3x + 4 \leq 19$.

[A] $2 \leq x \leq 5$

[B] $-2 \leq x \leq 5$

[C] $x \leq 5$

[D] $x \geq 2$

[E] $-5 \leq x \leq -2$

The answer is: A

The first equation solving step used to isolate the x is the subtraction of 4. In a conjunction, the subtraction, as well as the division of 3 following, must be performed on all three parts of the inequality.

$$10 \leq 3x + 4 \leq 19$$
$$6 \leq 3x \leq 15$$
$$2 \leq x \leq 5$$

20. Solve for x: $x^2 + 10x - 24 = 0$

[A] (−5,12)

[B] (−10,8)

[C] (12,2)

[D] (10,8)

[E] (−12,2)

The answer is: E

Factor the trinomial and set each factor equal to zero to solve for x.
$$x^2 + 10x - 24 = 0$$
$$(x+12)(x-2) = 0$$
$$x + 12 = 0 \text{ or } x - 2 = 0$$

21. **Find a quadratic equation with roots of 4 and -9.**

[A] $x^2 - 5x + 36 = 0$

[B] $x^2 + 5x - 36 = 0$

[C] $4x^2 - 9x - 5 = 0$

[D] $x^2 + 4x - 9 = 0$

[E] $5x^2 - 9x + 4 = 0$

The answer is: B

If r is a root of a polynomial, then (x − r) is a factor.
$$(x-4)(x+9) = 0$$
$$x^2 - 4x + 9x - 36 = 0$$
$$x^2 + 5x - 36 = 0$$

22. **Solve:** $4800 \leq 200x - 2x^2$

[A] $-40 \leq x \leq 40$

[B] $x \leq 40$

[C] $40 \leq x \leq 60$

[D] $x = 40$

[E] $x = -40$

The answer is: C

Start the solution process by setting the inequality less than zero.
$2x^2 - 200x + 4800 \leq 0$

One approach is to then graph the parabolic function on a calculator, and, after adjusting the window appropriately, find that the parabola is below the x axis, or less than zero, between 40 and 60.

Alternatively, factor and solve the inequality: $2x^2 - 200x + 4800 \leq 0$
$$x^2 - 100x + 2400 \leq 0$$
$$(x-40)(x-60) \leq 0$$

But this work indicates that 40 and 60 are boundaries to a solution interval. Values must be tested in order to determine the actual values where the inequality is less than zero.

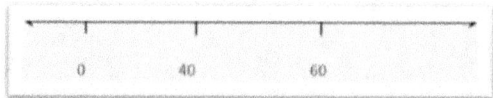

For instance, test a value greater than 60, like 70:

$2(70)^2 - 200(70) + 4800 = 600$ which is greater than zero. So the solution does not exist in this interval. Continue with the remaining intervals as shown on the number line to conclude that the polynomial is less than zero between 40 and 60.

23. Solve: $|3x+2| = 4x+5$

[A] $x = -3$

[B] $x = -1$

[C] $x = 3$

[D] $x = 1$

[E] $x = 6$

The answer is: B

Use the definition of absolute value to set up two equations.

3x + 2 = 4x + 5 or 3x + 2 = -4x - 5

-x = 3 or -7x = 7

x = -3 or x = -1

Check for extraneous solutions:

$|3x+2| = 4x+5$ | $|3x+2| = 4x+5$
$|3(-3)+2| ? 4(-3)+5$ | $|3(-1)+2| ? 4(-1)+5$
$|-9+2| ? -12+5$ | $|-3+2| ? -4+5$
$7 \neq -7$ | $1 = 1$

Therefore -3 is not a solution while -1 is.

24. Solve: $|3x - 5| = \dfrac{1}{2}$

[A] $x = -\dfrac{11}{6}, -\dfrac{3}{2}$

[B] $x = -\dfrac{11}{6}, \dfrac{3}{2}$

[C] $x = \dfrac{11}{6}, -\dfrac{3}{2}$

[D] $x = \dfrac{11}{6}, \dfrac{3}{2}$

[E] $x = 11, \dfrac{3}{2}$

The answer is: D

Use the definition of absolute value to set up two equations.

$$3x - 5 = \dfrac{1}{2} \qquad\qquad 3x - 5 = -\dfrac{1}{2}$$

$$3x = \dfrac{11}{2} \qquad\qquad 3x = \dfrac{9}{2}$$

$$x = \dfrac{11}{6} \qquad\qquad x = \dfrac{3}{2}$$

25. **Solve:** $2|3x+9| < 36$

[A] $x < -9$

[B] $x > 3$

[C] $3 < x < 9$

[D] $-9 < x < -3$

[E] $-9 < x < 3$

The answer is: E

First isolate the absolute value, then set up a conjunction to solve.

$$2|3x+9| < 36$$
$$|3x+9| < 18$$
$$-18 < 3x+9 < 18$$
$$-27 < 3x < 9$$
$$-9 < x < 3$$

26. Solve for x and y:

$4x + 3y = -1$
$5x + 4y = 1$

[A] $x = -7, y = 9$

[B] $x = 7, y = -9$

[C] $x = 7, y = 9$

[D] $x = -7, y = -9$

[E] $x = y = 7$

The answer is: A

Using the elimination method:

$4x + 3y = -1 \xrightarrow{-4} -16x - 12y = 4$
$5x + 4y = 1 \xrightarrow{3} 15x + 12y = 3$

After combining the two new equations, -x = 7 or x = -7. Substitute into one equation to find y.

4(-7) + 3y = -1, y = 9. Therefore the solution to the system is (-7, 9).

27. **Which point is in the solution set for the system of inequalities below?**

$$x - 7 < 1$$
$$y < 2x - 1$$

[A] $(-1,-1)$

[B] $(-2,-1)$

[C] $(0,1)$

[D] $(0,-2)$

[E] $(1,1)$

The answer is: D

Only point D satisfies both equations algebraically: $0 - 7 < 1$, $-2 < 2(0) - 1$. Additionally, a graph shows that only point D is within the shaded solution region. (Note that point E is on the line, which is not part of the solution region)

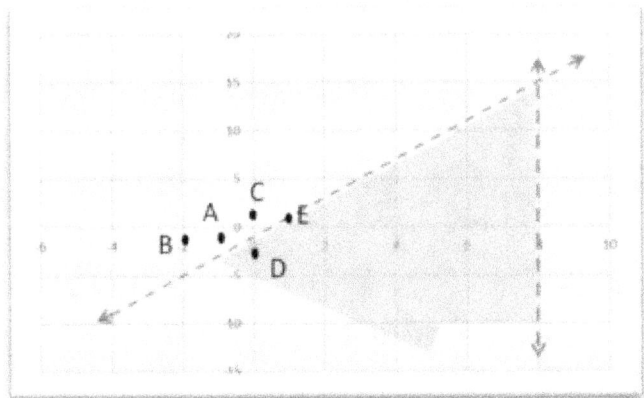

28. Solve: $3^{2x-1} = 27$

   [A] $x = 2$

   [B] $x = -3$

   [C] $x = -2$

   [D] $x = 3$

   [E] $x = \dfrac{2}{3}$

   The answer is: A

   Since $27 = 3^3$, the equation can be rewritten: $3^{2x-1} = 3^3$. With the same base on each side, the exponents must be equal. 2x – 1 = 3, so x = 2.

29. Solve: $\log_b(x^2) = \log_b(2x - 1)$

   [A] $x = -2$

   [B] $x = 1$

   [C] $x = -1$

   [D] $x = 2$

   [E] $x = 4$

   The answer is: B

   Since the log is of the same base on each side of the equation, the arguments are equal.
   $$x^2 = 2x - 1$$
   $$x^2 - 2x + 1 = 0$$
   (x - 1)(x - 1) = 0, x = 1

30. Solve: $\log_2(x) + \log_2(x-2) = 3$

[A] $x = 4$

[B] $x = -4, 2$

[C] $x = -4, -2$

[D] $x = 4, -2$

[E] $x = 2$

The answer is: A

Combine the log expressions into one using the product/sum rule:
$\log_b(a) + \log_b(c) = \log_b(ac)$

$$\log_2 x(x-2) = 3$$

Then rewrite the relationship exponentially.

$$2^3 = x(x-2)$$
$$8 = x^2 - 2x$$
$$0 = x^2 - 2x - 8$$
$$0 = (x-4)(x+2), x = 4, -2$$

However, when preparing to check the solutions, x cannot be a negative number as the log function is not defined over negative numbers.
Check the positive value for x.

$$\log_2(4) + \log_2(4-2) ? 3$$
$$\log_2(4) + \log_2(2) ? 3$$
$$2 + 1 = 3$$

31. If $f(x) = -3x + 8$, find

[A] 23

[B] -23

[C] 7

[D] -7

[E] 21

The answer is: D

Evaluate the function for x = 5. $f(5) = -3(5) + 8 = -15 + 8 = -7$

32. Find the zeros of the function $h(x) = \dfrac{x-9}{x+2}$.

[A] {9}

[B] {−2}

[C] $\left\{-\dfrac{9}{2}\right\}$

[D] {−2, 9}

[E] This function has no zeros.

The answer is: A

The zero of a function is defined as the (x) input required to give the function a (y) value of zero. This function will be zero when the numerator has a value of zero: $x - 9 = 0$, $x = 9$. When the denominator of this function equals zero, at x = -2, the function will be undefined.

33. Which number line shows the solution to $7x - 5 \geq 9x - 17$?

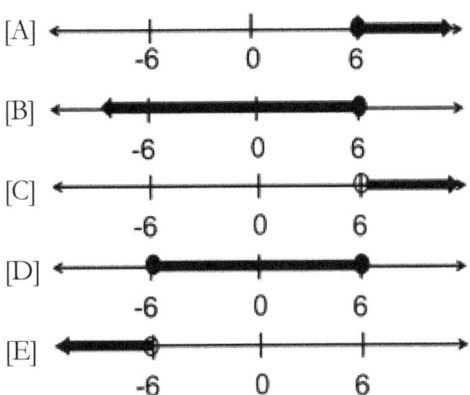

The answer is: B

First gather all the x terms on one side of the inequality and the numbers on the other.

$$7x - 5 \geq 9x - 17$$
$$-2x \geq -12$$

When dividing both sides of an inequality by a negative number, the inequality sign is reversed. So division by -2 on both sides results in $x \leq 6$ which is graphed in choice B.

34. Which graph represents the equation of $y = x^2 + 3x$?

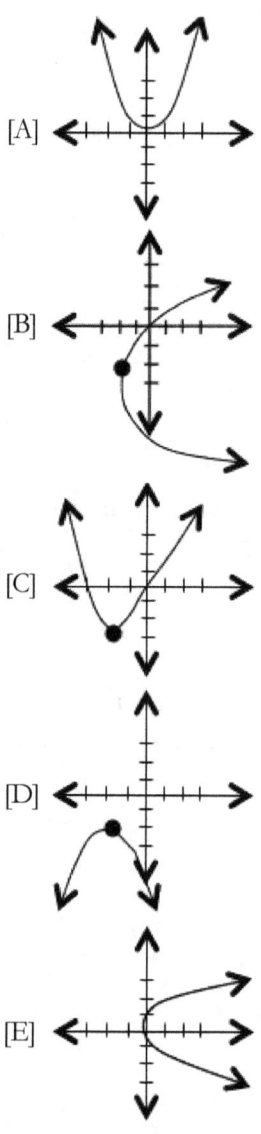

The answer is: C

Find the x intercepts of the graph by making y = 0 and solving for x.
$$0 = x^2 + 3x$$
$$0 = x(x+3), x = 0 \text{ or } x = -3$$
Therefore the x intercepts are (0, 0) and (0, -3) which are seen in the graph pictured in choice C.

35. Based on the given table, if $y_1 = x^3$, what is the equation for

x	-2	-1	0	1	2	3
y_1	-8	-1	0	1	8	27
y_2	-18	-11	-10	-9	-2	-17

[A] $y_2 = x^5$

[B] $y_2 = -x^3$

[C] $y_2 = (-x)^3$

[D] $y_2 = (x-10)^3$

[E] $y_2 = x^3 - 10$

The answer is: E

When comparing each y_2 to each y_1, a difference of 10 is observed, resulting in choice E.

36. Identify the domain and range of the relation:
$$\{(2,-5),(4,31),(11,-11),(-21,3)\}$$

[A] Domain is $\{-21\}$, range is $\{-11\}$.

[B] Domain is $\{-5,31,-11,3\}$, range is $\{2,4,11,-21\}$.

[C] Domain is $\{11\}$, and range is $\{31\}$.

[D] Domain and range are indeterminate.

[E] Domain is $\{2,4,11,-21\}$, range is $\{-5,31,-11,3\}$.

The answer is: E

In a set of ordered pairs, the domain is made up of the values for x, while the range consists of the y values.

37. Determine the domain of $y = -\sqrt{-2x + 3}$.

[A] $x = 3$

[B] $x \leq \dfrac{3}{2}$

[C] $x > \dfrac{3}{2}$

[D] $x = 2$

[E] $x = 0$

The answer is: B

In order to keep the function defined over the real numbers, the radicand must remain non-negative.

$-2x + 3 \geq 0$

$-2x \geq -3$

$x \leq \dfrac{3}{2}$ as the inequality is reversed when dividing by a negative number.

38. For the function $h(x)$ whose graph is shown below, select the domain and range.

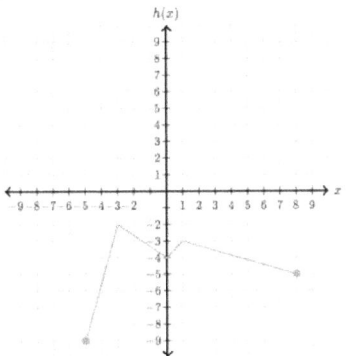

[A] Domain is $-5 \leq x \leq 8$, range is $-9 \leq h \leq -2$.

[B] Domain is -5, range is -5.

[C] Range is $-5 \leq x \leq 8$, domain is $-9 \leq h \leq 2$.

[D] Domain is $x \geq -5$, range is $h \geq -9$.

[E] Domain is 8, range is -2.

The answer is: A

Choice A represents the graphed x values for the domain, and the graphed y values for the range.

39. Given $f(x) = 3x^2 - 7x + 5$, find $f(-4)$.

[A] -71

[B] 25

[C] 81

[D] -25

[E] 71

The answer is: C

Evaluate the function for x = -4
$$f(-4) = 3(-4)^2 - 7(-4) + 5 = 3(16) + 28 + 5 = 81$$

40. For $h(x) = 3x^2 + ax - 1$, $h(3) = 8$, find the value of a.

[A] 6

[B] −6

[C] −18

[D] 18

[E] 27

The answer is: B

Evaluate the function for x = 3, then solve for a.

$$h(3) = 8 = 3(3)^2 + a(3) - 1$$
$$8 = 27 + 3a - 1$$
$$-18 = 3a$$
$$a = -6$$

41. Given $f(x) = 3x^2 - 7x + 5$, find $\dfrac{f(x+h) - f(x)}{h}$

[A] $7h$

[B] $6xh - 7$

[C] $6x + 3h - 7$

[D] $3x + 6h + 7$

[E] $5x$

The answer is: C

$$\frac{f(x+h) - f(x)}{h} = \frac{3(x+h)^2 - 7(x+h) + 5 - [3x^2 - 7x + 5]}{h}$$
$$= \frac{3x^2 + 6xh + 3h^2 - 7x - 7h + 5 - 3x^2 + 7x - 5}{h}$$
$$= \frac{6xh + 3h^2 - 7h}{h}$$
$$= 6x + 3h - 7$$

42. Find the x- and y- intercepts for $5x - 3y = 15$.

[A] $x = 0, y = 0$

[B] $x = -3, y = 5$

[C] $x = -1, y = 5$

[D] $x = -5, y = 3$

[E] $x = 3, y = -5$

The answer is: E

To find the x intercept, make y = 0: 5x – 3(0) =15, x = 3.
To find the y intercept, make x = 0: 5(0) – 3y = 15, y = -5

43. **Which of the figures is a reflection of the triangle shown?**

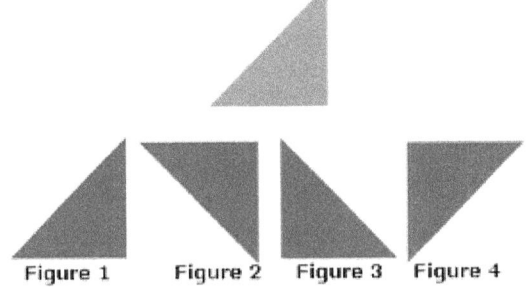

[A] Figure 1 and Figure 4

[B] Figure 4 and Figure 3

[C] Figure 2 and Figure 1

[D] Figures 2, 3 and 4

[E] Figure 1 and Figure 2

The answer is: D

Figure 2 represents a vertical flip, or a reflection over the x axis. Figure 3 represents a horizontal flip, or a reflection over the y axis. Figure 4 is a result of both of these reflections applied.

44. Name the transformation shown.

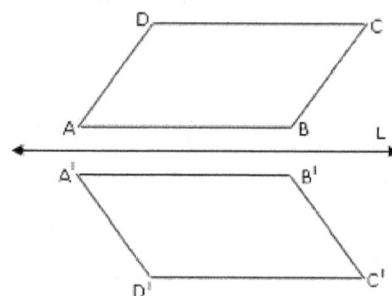

[A] Translation

[B] Rotation

[C] Reflection

[D] Dilation

[E] Cannot be determined

The answer is: C

The new image is created from a reflection over the x axis.

45. Find the inverse of $y = 3x - 2$.

[A] $y = \dfrac{1}{3x - 2}$

[B] $y = \dfrac{x + 2}{3}$

[C] $x = \dfrac{y + 2}{3}$

[D] $y - 3x - 2 = 0$

[E] $3x - y - 2 = 0$

The answer is: B

To find the inverse of a function, switch the x and y, then solve for the "new" y.

$x = 3y - 2$

$x + 2 = 3y$

$y = \frac{x+2}{3}$

46. Find the inverse of $f(x) = -\frac{1}{3}x + 1$

[A] $f^{-1}(x) = 1$

[B] $f^{-1}(x) = 3x$

[C] $f^{-1}(x) = 3x - 3$

[D] $f^{-1}(x) = -3x + 3$

[E] $f^{-1}(x) = x^2$

The answer is: D

Again, reverse the x and y, where f(x) = y.

$y = -\frac{1}{3}x + 1$

$x = -\frac{1}{3}y + 1$

$x - 1 = -\frac{1}{3}y$

$-3x + 3 = y$

Now the "new" y can be noted as $f^{(-1)}(x)$, which suggests the inverse of f(x).

47. If $f(x) = 3x - 2$ and $g(x) = \dfrac{x}{3} + \dfrac{2}{3}$, which of the following is true:

   [A] $f(x)$ is the inverse of $g(x)$.

   [B] $f(x) = g^{-1}(x)$

   [C] There is no connection between $f(x)$ and $g(x)$.

   [D] $g(x) = f(x)$

   [E] A and B

   The answer is: E

   Choice A and B each communicate that the two functions are inverses of each other. To prove this, derive one from the other, or show that $f(g(x)) = x = g(f(x))$

   $$f(g(x)) = 3\left(\dfrac{x}{3} + \dfrac{2}{3}\right) - 2 = x + 2 - 2 = x$$

   $$g(f(x)) = \dfrac{3x-2}{3} + \dfrac{2}{3} = \dfrac{3x}{3} = x$$

48. Identify the real numbers in the list: $1.67, \pi, \sqrt{5}, 0$

   [A] All

   [B] $1.67, \sqrt{5}, 0$

   [C] $1.67, 0$

   [D] 0

   [E] None

   The answer is: A

   All of the given numbers are real. (None are imaginary)

49. Which of the following is false?

   [A] Every rational number is a real number.

   [B] Every imaginary number is a real number.

   [C] Every integer is a whole number.

   [D] Every integer is a real number.

   [E] Every natural number is positive.

   The answer is: E

   Choice B is false because no imaginary numbers are real. Choice C is false because some integers are negative, while whole numbers are zero or positive.

50. Which selection below is NOT a real number?

   [A] -3

   [B] $0.6666...$

   [C] $\dfrac{\pi}{2}$

   [D] $3+\sqrt{2}$

   [E] $3i$

   The answer is: E

   Choice E shows an imaginary number which is not real.

51. Simplify $\sqrt{-9}$.

   [A] -3

   [B] -3i

   [C] 3i

   [D] 3

   [E] 0

   The answer is: C
   $\sqrt{-9} = \sqrt{9} \cdot \sqrt{-1} = 3\text{i}$

52. Simplify $(i)(2i)(-3i)$.

   [A] 6i

   [B] -6i^3

   [C] -6i

   [D] 0

   [E] 6i^3

   The answer is: A
   $(i)(2i)(-3i) = -6i^3 = -6(-i) = 6i$

53. Simplify i^{17}.

   [A] 17i

   [B] i

   [C] -17i

   [D] -i

   [E] 1

   The answer is: B
   $i^{17} = (i^4)^4 \cdot i = 1i = i$

54. List the first four terms of the following sequence, beginning with $n = 0$.

$$A_n = \frac{(-1)^n}{(n+1)!}$$

[A] $\frac{1}{2}, 1, \frac{3}{2}, 2$

[B] $-1, -\frac{1}{2}, 0, \frac{1}{2}$

[C] $0, 1, 2, 3$

[D] $1, -\frac{1}{2}, \frac{1}{6}, -\frac{1}{24}$

[E] $0, -1, -\frac{1}{2}, \frac{2}{3}$

The answer is: D

Evaluate the sequence rule for n = 0, 1, 2, 3

$$\frac{(-1)^0}{(0+1)!}, \frac{(-1)^1}{(1+1)!}, \frac{(-1)^2}{(2+1)!}, \frac{(-1)^3}{(3+1)!}$$

$$\frac{1}{(1)!}, \frac{-1}{(2)!}, \frac{1}{(3)!}, \frac{-1}{(4)!}$$

$$1, \frac{-1}{4}, \frac{1}{6}, \frac{-1}{24}$$

55. Expand the following series and find the sum:

$$\sum_{n=0}^{4} 2n$$

[A] 20

[B] 8

[C] 16

[D] 4

[E] 32

The answer is: A

The five terms in the expansion are 0 + 2 + 4 + 6 + 8 so the sum is 20.

56. Write the series in sigma notation: $-3+0+9+24+45+72+105$

[A] $\sum_{a=0}^{6} 3a^2$

[B] $\sum_{a=0}^{6} 3a^2 - 3$

[C] $\sum_{a=0}^{6} a^2 - 3$

[D] $\sum_{a=1}^{6} 3a^2 - 1$

[E] $\sum_{a=0}^{5} a^2 - 3$

The answer is: B

Working backwards, substituting the start value into the rule for each sum, eliminates all but choice B. That is, choice B is the only rule where a_0 or a_1 matches the first listed term in the sum. Furthermore, evaluating the rule for n = 0 – 6 proves that all 7 terms match confirming that B is the correct choice.

57. Find $\dfrac{8!}{6!2!}$

[A] $\dfrac{2}{3}$

[B] $\dfrac{4}{6}$

[C] 28

[D] 48

[E] 24

The answer is: C

Most calculators can handle factorials of this size, however it is practical to know how to simplify an expression of this sort.

$$\dfrac{8!}{6!2!} = \dfrac{8 \cdot 7 \cdot 6!}{2 \cdot 6!} = \dfrac{8}{2} \cdot 7 = 4 \cdot 7 = 28$$

58. Expand the binomial $(2x+3y)^4$

[A] $16x^4 + 24x^3y + 36x^2y^2 + 54xy^3 + 81y^4$
    $2x^4 + 6x^3y + 6x^2y^2 + 6xy^3 + 3y^4$

[B] $16x^4 + 81y^4$

[C] 16x4 + 96x3y + 216x2y2 + 216xy3 + 81y4

[D] $16x^4 + 24x^3y^3 + 36x^2y^2 + 54xy + 81y^4$

[E] $x^4 + 4x^3y + 6x^2y^2 + 4xy^3 + y^4$

The answer is: C

This problem can be expanded by binomial and trinomial multiplication, using coefficients of Pascal's Triangle, or following the Binomial Expansion Formula.

$(2x+3y)^4 = (2x+3y)^2 \cdot (2x+3y)^2$
$= (4x^2 + 12xy + 9y^2)(4x^2 + 12xy + 9y^2)$
$= 16x^4 + 48x^3y + 36x^2y^2 + 48x^3y + 144x^2y^2 + 108xy^3 + 36x^2y^2 + 108xy^3 + 81y^4$
$= 16x^4 + 96x^3y + 216x^2y^2 + 216xy^3 + 81y^4$

Check the third term using the Binomial Expansion Formula:

$$(a+b)^n = \sum_{k=0}^{n} {}_nC_k a^{n-k} b^k$$

To find the third term, k = 2. (since the formula starts with k = 0)

$(2x+3y)^4 : \quad {}_4C_2 (2x)^{4-2}(3y)^2 = 6 \cdot 4x^2 \cdot 9y^2 = 216x^2y^2$

59. Evaluate the determinant of the matrix:
$$\begin{pmatrix} -2 & 4 \\ -4 & 3 \end{pmatrix}$$

[A] 10

[B] -24

[C] 4

[D] -10

[E] 24

The answer is: A

Given the matrix $\begin{pmatrix} a & b \\ c & d \end{pmatrix}$, the determinant can be calculated by $ad - cb$

(-2)(3) − (-4)(4) = -6 − (-16) = -6 + 16 = 10

60. Evaluate the determinant of the matrix for $y = 4$.
$$\begin{pmatrix} -5y & 3y \\ y-1 & y-3 \end{pmatrix}$$

[A] 35

[B] 12

[C] -56

[D] -12

[E] 56

The answer is: C

First evaluate the matrix for y = 4. $\begin{pmatrix} -5(4) & 3(4) \\ 4-1 & 4-3 \end{pmatrix} = \begin{pmatrix} -20 & 12 \\ 3 & 1 \end{pmatrix}$

Then use the determinant formula described above: -20(1) − 3(12) = -56

CLEP College Algebra
Sample Test 2

Sample Test Two

Select the best answer for each question below.

1. Simplify $(2-3i)(4i)$.

    [A] $-12+8i$

    [B] $12+8i$

    [C] $8-12i$

    [D] $8+12i$

    [E] $-8+12i$

2. Solve $\sqrt{(n^2+16)} = 3n$

    [A] 2

    [B] ± 2

    [C] $\pm\sqrt{2}$

    [D] $\pm\dfrac{4}{3}$

    [E] No Real Solution

3. Insert mathematical symbols to make the given calculation correct. $3+5\cdot 4-1=18$

    [A] Place parenthesis around 5•4

    [B] Place parenthesis around 3+5

    [C] Place parenthesis around 4 – 1

    [D] Change the subtraction to addition of the opposite

    [E] No symbols are needed. The calculation is already correct.

4. Find the solution to the system of equations.
$$\begin{cases} 4x+2y=18 \\ y=-2x+9 \end{cases}$$

   [A] No solution

   [B] Infinitely many solutions

   [C] (2, 1)

   [D] (9, 18)

   [E] (0, 0)

5. If $h(x) = \dfrac{3x+4}{x}$ for all real values of $x \neq 0$, find $h^{-1}(x)$.

   [A] $h^{-1}(x) = \dfrac{-3x-4}{x}$

   [B] $h^{-1}(x) = \dfrac{x}{3x+4}$

   [C] $h^{-1}(x) = \dfrac{4}{x-3}$

   [D] $h^{-1}(x) = \left(\dfrac{1}{x}\right)(3x+4)$

   [E] None of the above represent $h^{-1}(x)$.

6. What values of x will ensure that all values of this function are real numbers? $f(x) = \sqrt{(3-2x)}$

   [A] $\left\{ x \mid x \leq \dfrac{3}{2} \right\}$

   [B] $\left\{ x \mid x \geq \dfrac{2}{3} \right\}$

   [C] $\left\{ x \mid x - \dfrac{3}{2} \leq x \leq \dfrac{3}{2} \right\}$

   [D] $\left\{ x \mid x \neq \dfrac{3}{2} \right\}$

   [E] All real numbers, x

7. Which of the following would produce an odd number?

   [A] $(Odd + Even + Odd) \times Odd$

   [B] $(Odd \times Odd) + Odd$

   [C] $(Even + Odd + Even) \times Odd$

   [D] $(Odd + Odd) \times (Even + Odd)$

   [E] $(Odd \times Odd) - Odd$

8. Based on the given table, if $y_1 = x^3$, what is the equation for y_2?

x	-2	-1	0	1	2	3
y_1	-8	-1	0	1	8	27
y_2	8	1	0	-1	-8	-27

   [A] $y_2 = x^5$

   [B] $y_2 = -x^3$

   [C] $y_2 = (-x)^3$

   [D] $y_2 = (x - 10)^3$

   [E] $y_2 = x^3 - 10$

9. What is the determinant of the matrix below?

   $$\begin{bmatrix} 7 & 5 \\ 1 & 4 \end{bmatrix}$$

   [A] -5

   [B] -23

   [C] 5

   [D] 23

   [E] 33

10. Find the zeros of the function
$$h(x) = \frac{2x+6}{x+2}.$$

[A] $\{-3\}$

[B] $\{-2\}$

[C] $\{-6\}$

[D] $\{-2, -3\}$

[E] This function has no zeros.

11. Solve for x. $\dfrac{4}{x+2} = \dfrac{3}{x-7}$

[A] -1

[B] 0

[C] 7

[D] 31

[E] No solution

12. Which of the equations below, when graphed with $y = 10^x$, will show a reflection over the line $= x$?

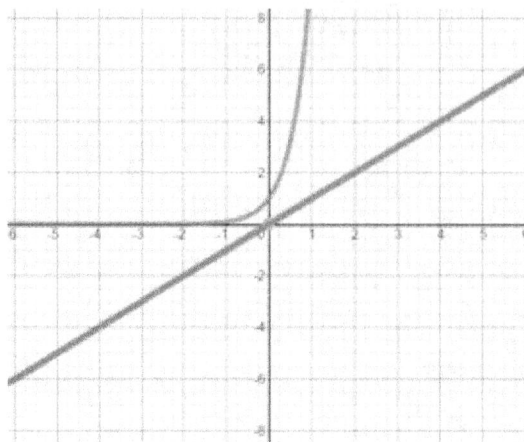

[A] $y = (-10)^x$

[B] $y = -10^x$

[C] $y = 10^{-x}$

[D] $y = \log x$

[E] $y = \ln x$

13. If a certain bacteria grows at a rate according to the function $b(t) = 50(1.5)^{0.3t}$ where t is measured in days and $b(0) = 50$ milligrams, find the amount of bacteria present after 4 days.

[A] 54

[B] 81

[C] 178

[D] 6,487

[E] 20,000

14. Given the function $f(x)=\sqrt{x}-10$ which translation to the function shown in the choices below would ensure the new range values remained greater than zero?

   [A] $-f(x)$

   [B] $f(-x)$

   [C] $f(x) + 10$

   [D] $f(x) + 11$

   [E] None of the above

15. Find the product of the two matrices. $\begin{bmatrix} 2 & 1 \\ 0 & 2 \end{bmatrix} \times \begin{bmatrix} 1 & 1 \\ 2 & 0 \end{bmatrix}$

   [A] $\begin{bmatrix} 4 & 2 \\ 4 & 0 \end{bmatrix}$

   [B] $\begin{bmatrix} 2 & 1 \\ 0 & 0 \end{bmatrix}$

   [C] $\begin{bmatrix} 4 & 0 \\ 0 & 0 \end{bmatrix}$

   [D] $\begin{bmatrix} 6 & 3 \\ 4 & 2 \end{bmatrix}$

   [E] $\begin{bmatrix} 3 & 2 \\ 2 & 2 \end{bmatrix}$

16. Find the point(s) of intersection of the graphs $f(x) = 4x^2 + 8$ and $g(x) = x^3 + 2x$.

   [A] (0, 0) and (0, 8)

   [B] (0, 8)

   [C] $(2\sqrt{2}, 16)$

   [D] (4, 72)

   [E] There is no point of intersection.

17. Find the equation of the line that passes through the point (3, 7) and has a slope of $\frac{1}{3}$.

   [A] $y = 3x + 7$

   [B] $y = \frac{1}{3}x + 7$

   [C] $y = \frac{1}{3}x + 6$

   [D] $x - 3y + 18 = 0$

   [E] Both C and D

18. Find the zeros of the function $f(x) = x2 - 12x - 13$.

   [A] -1

   [B] 0

   [C] 12

   [D] 13

   [E] Both A and D

19. If $h(x) = 3x - 7$, find $-h(x)$.

   [A] $-h(x) = 3x + 7$

   [B] $-h(x) = 7 - 3x$

   [C] $-h(x) = 7x - 3$

   [D] $-h(x) = -10$

   [E] None of the above

20. Two friends post a video on the Internet on Monday. On Tuesday each of them forwards the video to three friends. On Wednesday, each of those friends forwards the video to three more friends. The same thing happens Thursday and Friday. How many people have seen the video by the end of Friday?

   [A] 121

   [B] 162.

   [C] 240.

   [D] 242

   [E] 728

21. Factor completely $x4 - 4x^2 - 45$

   [A.] $(x2 - 15)(x2 + 3)$

   [B] $(x2 + 9)(x2 - 5)$

   [C] $(x2 + 5)(x + 3)(x - 3)$

   [D] $(x + 5)(x - 5)(x + 3)(x - 3)$

   [E] None of the above

22. The fine for an overdue library book is 50 cents on the first overdue day and increases by 5 cents on each subsequent day. How great a fine is due for a book that has been overdue for 6 days?

    [A] $0.30

    [B] $0.75

    [C] $0.80

    [D] $3.00

    [E] $3.75

23. Which function below is not continuous over the set of real numbers?

    [A] $f(x) = |x|$

    [B] $g(x) = x^2$

    [C] $h(x) = \dfrac{1}{x}$

    [D] $q(x) = \begin{cases} 0 & \text{for } x > 10 \\ \sqrt{10-x} & \text{for } x \leq 10 \end{cases}$

    [E] Both C and D are not continuous.

24. Find a polynomial function with zeros at -3, $-\sqrt{2}$, 3, $\sqrt{2}$.

    [A] $f(x) = 3x^4 - 3x^3 + (\sqrt{2})x^2 - (\sqrt{2})x$

    [B] $g(x) = x^2(x-3) + x^3(x-\sqrt{2})$

    [C] $h(x) = (x^2+9)(x^2+2)$

    [D] $p(x) = x^4 - 11x^2 + 18$

    [E] $t(x) = x^4 - 3x^2 + \sqrt{2}$

158   CLEP Algebra

25. If **Pv = nrt**, solve for n.

 [A] $n = \frac{Pv}{rt}$

 [B] n = Pv − rt

 [C] n = Pvrt

 [D] n = P + v − r − t

 [E] none of the above

26. Which of the following expressions below is equivalent to $(x-7)^2$?

 [A] $x^2 + 49$

 [B] $x^2 - 49$

 [C] $x^2 - 14x + 49$

 [D] $49x^2$

 [E] $2x - 14$

27. Solve the equation $\frac{5a - 10x}{3} = \frac{1}{5}$

 [A] {47, 53}

 [B] $\left\{\frac{47}{25}, \frac{53}{25}\right\}$

 [C] $-\frac{47}{5}$

 [D] −47

 [E] 53

28. Simplify $\dfrac{5a-10x}{3} = \dfrac{1}{5}$

[A] $\dfrac{x^2 + 3x + 1}{x}$

[B] $\dfrac{x^2 + 11x + 4}{2x + 3}$

[C] $\dfrac{x^2 + 11x + 9}{3x}$

[D] $2x + 4$

[E] 3

29. Select the statement below that explains the best first step to solving the equation $\dfrac{3x - 7}{4} = 5x + 1$

[A] Add 7 to both sides.

[B] Subtract $3x$ from both sides.

[C] Divide both sides by 5.

[D] Multiply both sides by $\dfrac{1}{3}$.

[E] Multiply both sides by 4.

30. Which number line shows the solution to $7x + 5 > 9x + 17$?

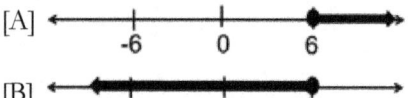

31. Subtract $3-2i$ from $10-7i$

   [A] $16i$

   [B] $7 + 5i$

   [C] $7 + 9i$

   [D] $13 + 5i$

   [E] $-7 - 5i$

32. Given $g(x) = 2x^2 + 4$, find $g(-3)$.

   [A] -14

   [B] -8

   [C] 16

   [D] 22

   [E] 40

33. If $g(x) = x^2 + 9$ and $f(x) = x^2$, find $f(g(x))$.

   [A] $x + 3$

   [B] $2x^2 + 9$

   [C] $x^4 + 9$

   [D] $x^4 + 81$

   [E] $x^4 + 18x^2 + 81$

34. Which of the following equations does not represent a function?

[A] $y = x^2$

[B] $x = y$

[C] $x = y - 5$

[D] $x = 8$

[E] $y = 10$

35. A canoe leaves a dock and paddles north across a river at 1.5 mph while the river current carries the canoe eastward at 2 mph. How far is the canoe from the dock after 30 minutes?

[A] 0.75 mi

[B] 1 mi

[C] 1.25 mi

[D] 1.5 mi

[E] 2 mi

36. Find the equation of a line that passes through the origin, and is parallel to the line $2x + 7y = 14$

[A] $y = -\frac{2}{7}x$

[B] $y = \frac{7}{2}x$

[C] $y = x$

[D] $x + y = 14$

[E] $x + y = 0$

37. In the given graph, if $y_1 = x^2$, then which is the most likely equation to represent y^2?

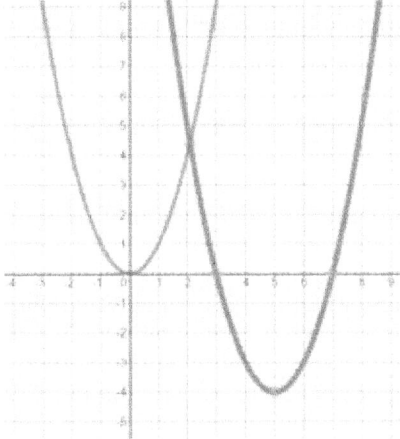

[A] $y_2 = 9x^2$

[B] $y_2 + 4 = x^2 + 5$

[C] $y_2 = (x-5)^2 - 4$

[D] y2 = (x - 5)(x - 4)

[E] None of the above

38. Given the piecewise function $f(x) = \begin{cases} x+3 & \text{for } x \geq 0 \\ 5 & \text{for } x > 0 \end{cases}$ find $f(8)$.

[A] 2

[B] 5

[C] 8

[D] 11

[E] None of the above

39. On a certain standardized test, Student A scores 50 on math, 60 on reading, and 48 on writing. Student B scores 48 on math, 45 on reading, and 55 on writing. Which matrix below best represents this data?

[A] $\begin{bmatrix} A & 50 \\ B & 48 \end{bmatrix}$

[B] $\begin{bmatrix} 50 & 60 \\ 45 & 55 \end{bmatrix}$

[C] $\begin{bmatrix} 50 & 60 & 48 \\ 48 & 45 & 55 \end{bmatrix}$

[D] $\begin{bmatrix} 50 & 60 & 48 \\ 55 & 45 & 48 \end{bmatrix}$

[E] $\begin{bmatrix} 50 & 60 & 48 \\ 75 & 60 & 50 \\ 45 & 48 & 55 \end{bmatrix}$

40. Given the equation of a parabola, $y = x^2$, which equation below represents a transformation best described as a shift of 10 units up and 3 units to the left?

[A] $y = 3x^2 + 10$

[B] $y = 10(x + 3)2$

[C] $y = (x + 3)^2 + 10$

[D] $y = (x - 3)^2 - 10$

[E] $y = (x - 3)^2 + 10$

41. Find the area of the region bounded by $\begin{cases} x^2 + y^2 = 25 \\ x \geq 0 \\ y \geq 0 \end{cases}$

    [A] 50π

    [B] 25π

    [C] $\dfrac{25\pi}{4}$

    [D] $\dfrac{5\pi}{4}$

    [E] 25

42. The ages of the participants in a hula-hoop contest are as follows:
    10, 18, 22, 17, 77, 19, 13, 20, 10, 15. Which measure would most accurately represent the data as a whole?

    [A] range

    [B] interquartile range

    [C] mode

    [D] mean

    [E] median

43. Which equation below represents a parabola with axis of symmetry x = 3?

    [A] $y^2 = (x - 3)$

    [B] $y = (x - 3)^2$

    [C] $x^2 + y^2 = 9$

    [D] $y = x^2 + 3$

    [E] $y = 3x^2$

44. Expand $(x+2)^3$

   [A] $6x$

   [B] $3x + 6$

   [C] $x^3 + 8$

   [D] $x^3 + 3x^2 + 8$

   [E] $x^3 + 6x^2 + 12x + 8$

45. The interquartile range shown in the boxplot is

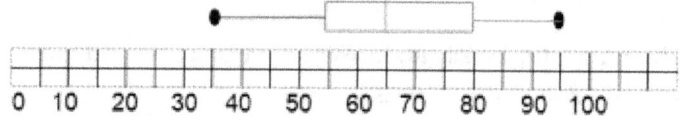

   [A] 25

   [B] 30

   [C] 35

   [D] 60

   [E] 95

46. State the domain of the function $f(x) = \dfrac{3x-6}{x^2-25}$

   [A] $x \neq 2$

   [B] $x \neq 5, -5$

   [C] $x \neq 2, -2$

   [D] $x \neq 5$

   [E] All real numbers

47. Given the function $f(x) = \begin{cases} x^2 + 5 & \text{for } x \geq 0 \\ x + 5 & \text{for } x > 0 \end{cases}$ for what values of x is the graph increasing?

   [A] For x > 0

   [B] For x < 0

   [C] For all real values of x

   [D] For x > 5

   [E] The graph is only decreasing.

48. Given the following table of values, what are possible equations for are y_1 and y_2?

x	y_1	y_2
-3	-3	-3
-2	-2	-2
-1	-1	-1
0	0	0
1	1	-1
2	2	-2
3	3	-3

   [A] , $y_2 = -x$

   [B] $y_1 = x, y_2 = |x|$

   [C] $y_1 = x, y_2 = -|x|$

   [D] $y_1 = -x, y_2 = |x|$

   [E] $y_1 = |x|, y_2 = -|x|$

49. In a card game, you get another turn if the card you draw is red or if it is a jack, queen, king, or ace. You are the first to draw from a full deck of 52 cards. What is the chance you will get another turn?

   [A] $\dfrac{2}{13}$

   [B] $\dfrac{4}{13}$

   [C] $\dfrac{1}{2}$

   [D] $\dfrac{17}{26}$

   [E] $\dfrac{21}{26}$

50. Find the 3rd term of the polynomial resulting from the expansion (presented in descending order) of $(x + 2)^5$

   [A] $7x$

   [B] $8x^3$

   [C] $40x^3$

   [D] $32x^5$

   [E] $80x^2$

51. Which of the following does not represent a real number?

   [A] $|14|$

   [B] $\sqrt{26}$

   [C] $\sqrt{26}$

   [D] $\dfrac{10}{5-5}$

   [E] Both C and D

52. Simplify $\dfrac{14x^5y^5}{7x^{-2}y}$

   [A] $2x^7y^4$

   [B] $7x^3y^5$

   [C] $21x^7y^6$

   [D] $\dfrac{1}{2x^7y^6}$

   [E] $2xy$

53. Simplify the imaginary expression $(3i^3)^3$.

   [A] $9i^6$

   [B] $9i^9$

   [C] $27i$

   [D] $27i^9$

   [E] -27

54. Which expression below is equivalent to $\log_6(25)$?

   [A] $\log(19)$

   [B] $\log(31)$

   [C] $\log_{12}(50)$

   [D] $\log_{25}(6)$

   [E] $2\log_6(5)$

55. Simplify $\dfrac{p!}{(p+2)!}$

   [A] $\dfrac{1}{2}$

   [B] $\dfrac{1}{p}$

   [C] $\dfrac{1}{p+2}$

   [D] $\dfrac{1}{(p+1)(p+2)}$

   [E] $(p+1)(p+2)$

56. Solve $|2x+7| < 19$

   [A] $-19 < x < 19$

   [B] $-13 < x < 6$

   [C] $x > -13$ or $x > 6$

   [D] $x < -13$ or $x < 6$

   [E] $x < -19$ or $x > 19$

57. If $3^x = 15$, then which of the following is also true?

   [A] $12^x = 60$

   [B] $9^x = 225$

   [C] $x = 5$

   [D] $x < 5$

   [E] $x < 0$

58. Solve for $x > 0$: $\log_4(x) = 16$.

    [A] 2

    [B] 4

    [C] 16

    [D] 64

    [E] 4,294,967,296

59. Find the rule that generates the following sequence: 3, 15, 75, 375,…

    [A] $a_n = 5n$

    [B] $a_n = 3(5n)$

    [C] $a_n = 3(5)^{n-1}$

    [D] $a_n = 3 + 5^n$

    [E] $a_n = 5^{n-1}$

60. Find the y intercept for the function $f(x) = x^2 - 4x - 12$

    [A] -12

    [B] -6

    [C] -2

    [D] 0

    [E] 12

SAMPLE TEST 2 ANSWER KEY

Question Number	Correct Answer	Your Answer
1	B	
2	C	
3	C	
4	B	
5	C	
6	A	
7	C	
8	E	
9	D	
10	A	
11	D	
12	D	
13	B	
14	D	
15	A	
16	D	
17	E	
18	E	
19	B	
20	D	
21	C	
22	B	
23	C	
24	D	
25	A	
26	C	
27	B	
28	A	
29	E	
30	E	

Question Number	Correct Answer	Your Answer
31	C	
32	D	
33	E	
34	D	
35	C	
36	A	
37	C	
38	D	
39	C	
40	C	
41	C	
42	E	
43	B	
44	E	
45	A	
46	B	
47	C	
48	C	
49	D	
50	C	
51	E	
52	A	
53	C	
54	E	
55	D	
56	B	
57	D	
58	E	
59	C	
60	A	

Sample Test 2 Explanations

1. Simplify $(2-3i)(4i)$.

   [A] $-12+8i$

   [B] $12+8i$

   [C] $8-12i$

   [D] $8+12i$

   [E] $-8+12i$

   The answer is: B

   Using the Distributive Property, we get $(2 \times 4i) - (3i \times 4i) = 8i - (12i2)$
   $= 8i - (-1)(12) = 12 + 8i$.

2. Solve $\sqrt{(n^2+16)} = 3n$

   [A] 2

   [B] ± 2

   [C] $\pm\sqrt{2}$

   [D] $\pm\dfrac{4}{3}$

   [E] No Real Solution

   The answer is: C

   Square both sides of the equation. $n^2 + 16 = 9n^2$
   $$16 = 8n^2$$
   $$2 = n^2$$

   Take the plus or minus square root of both sides $\pm\sqrt{2} = n$

3. **Insert mathematical symbols to make the given calculation correct.** $3+5\cdot 4-1=18$

   [A] Place parenthesis around 5•4

   [B] Place parenthesis around 3+5

   [C] Place parenthesis around 4 – 1

   [D] Change the subtraction to addition of the opposite

   [E] No symbols are needed. The calculation is already correct.

   The answer is: C

   Without inserting any symbols, the problem's answer is 22 as multiplication must be completed before the addition and subtraction. Choice C turns the calculation into

   3+5•(4-1)

   3+5•3

   3+15

   18

4. **Find the solution to the system of equations.**
   $$\begin{cases} 4x+2y=18 \\ y=-2x+9 \end{cases}$$

   [A] No solution

   [B] Infinitely many solutions

   [C] (2, 1)

   [D] (9, 18)

   [E] (0, 0)

   The answer is: B

   These two equations represent the same line, as the first can be algebraically manipulated to match the second: $4x+2y=18$
   $2y = -4x+18$
   $y = -2x+9$

   Since they are the same line, their intersection, or solution point, is every point on the infinitely long line, making choice B the correct answer.

5. If $h(x) = \dfrac{3x+4}{x}$ for all real values of $x \neq 0$, find $h^{-1}(x)$.

[A] $h^{-1}(x) = \dfrac{-3x-4}{x}$

[B] $h^{-1}(x) = \dfrac{x}{3x+4}$

[C] $h^{-1}(x) = \dfrac{4}{x-3}$

[D] $h^{-1}(x) = \left(\dfrac{1}{x}\right)(3x+4)$

[E] None of the above represent $h^{-1}(x)$.

The answer is: C

Start with $y = \dfrac{3x+4}{x}$ and replace x and y to find the inverse. $x = \dfrac{3y+4}{y}$

Then solve for y: $\quad x = \dfrac{3y+4}{y}$

Cross multiply: $\quad xy = 3y + 4$

Put y-terms on same side: $\quad xy - 3y = 4$

Factor out a y: $\quad y(x - 3) = 4$

Divide both sides by $(x - 3)$: $\quad y = \dfrac{4}{x-3}$

6. What values of x will ensure that all values of this function are real numbers? $f(x)=\sqrt{(3-2x)}$

[A] $\left\{x \mid x \leq \dfrac{3}{2}\right\}$

[B] $\left\{x \mid x \geq \dfrac{2}{3}\right\}$

[C] $\left\{x \mid x - \dfrac{3}{2} \leq x \leq \dfrac{3}{2}\right\}$

[D] $\left\{x \mid x \neq \dfrac{3}{2}\right\}$

[E] All real numbers, x

The answer is: A

To keep the function defined over the real numbers, the radicand must not be negative. Algebraically: $3-2x \geq 0$. Solving this inequality yields $x \leq \dfrac{3}{2}$, or choice A.

7. Which of the following would produce an odd number?

[A] $(Odd + Even + Odd) \times Odd$

[B] $(Odd \times Odd) + Odd$

[C] $(Even + Odd + Even) \times Odd$

[D] $(Odd + Odd) \times (Even + Odd)$

[E] $(Odd \times Odd) - Odd$

The answer is: C

Answers A and D fail because the quantity inside the opening parentheses is even. Even times odd is even. Answers B and E fail because Odd + Odd or Odd − Odd is an even number. Answer C represents an odd number (in parentheses) multiplied by an odd number.

8. Based on the given table, if $y_1 = x^3$, what is the equation for y_2?

x	-2	-1	0	1	2	3
y_1	-8	-1	0	1	8	27
y_2	8	1	0	-1	-8	-27

[A] $y_2 = x^5$

[B] $y_2 = -x^3$

[C] $y_2 = (-x)^3$

[D] $y_2 = (x - 10)^3$

[E] $y_2 = x^3 - 10$

The answer is: E

Each value of y_2 is the opposite of the value for y_1.

9. What is the determinant of the matrix below?
$$\begin{bmatrix} 7 & 5 \\ 1 & 4 \end{bmatrix}$$

[A] -5

[B] -23

[C] 5

[D] 23

[E] 33

The answer is: D

The determinant of a matrix $\begin{pmatrix} a & b \\ c & d \end{pmatrix}$ is $ad - bc = 7 \times 4 - 5 \times 1 = 23$.

10. Find the zeros of the function
$$h(x) = \frac{2x+6}{x+2}.$$

[A] {-3}

[B] {−2}

[C] {−6}

[D] {−2, -3}

[E] This function has no zeros.

The answer is: A
When $x = -3$, $h(x) = \frac{0}{-1} = 0$. −2 is not a zero but a value for which the function is undefined.

11. Solve for x. $\frac{4}{x+2} = \frac{3}{x-7}$

[A] -1

[B] 0

[C] 7

[D] 31

[E] No solution

The answer is: D
Start the solution process with cross multiplication.
$4(x-7) = 3(x+1)$
$4x - 28 = 3x + 3$
$x = 31$

12. Which of the equations below, when graphed with $y = 10^x$, will show a reflection over the line $= x$?

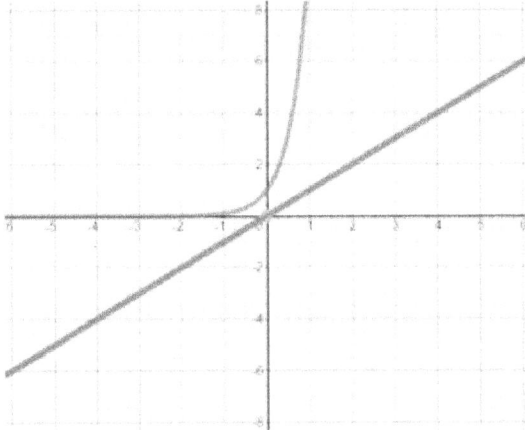

[A] $y = (-10)^x$

[B] $y = -10^x$

[C] $y = 10^{-x}$

[D] $y = \log x$

[E] $y = \ln x$

The answer is: D

The function that will be reflected over $x = y$ is the inverse function. $y = \log x$ is the inverse function of $y = 10^x$.

13. If a certain bacteria grows at a rate according to the function b(t) = 50(1.5)$^{0.3t}$ where t is measured in days and b(0) = 50 milligrams, find the amount of bacteria present after 4 days.

[A] 54

[B] 81

[C] 178

[D] 6,487

[E] 20,000

The answer is: B

Evaluate the function for t = 4: b(4) = 50(1.5)$^{0.3(4)}$ ≈ 50(1.6267) ≈ 81

14. Given the function $f(x) = \sqrt{x} - 10$ which translation to the function shown in the choices below would ensure the new range values remained greater than zero?

[A] -f(x)

[B] f(-x)

[C] f(x) + 10

[D] f(x) + 11

[E] None of the above

The answer is: D

The location of the minimum of $f(x) = \sqrt{x} - 10$ is (0, -10). Choice C would raise this point only up to (0,0). Therefore choice D, which raises the minimum to (0,1), is the translation that would keep range values positive. Be advised that A is not a valid choice as the x intercept of the original f(x) graph is (100,0). This point, while far outside the standard viewing range, represents the point where f(x) switches from negative values to positive. Choice A still contains this point. In other words, -f(x) Is a graph that does not have a range completely greater than zero.

15. Find the product of the two matrices. $\begin{bmatrix} 2 & 1 \\ 0 & 2 \end{bmatrix} \times \begin{bmatrix} 1 & 1 \\ 2 & 0 \end{bmatrix}$

[A] $\begin{bmatrix} 4 & 2 \\ 4 & 0 \end{bmatrix}$

[B] $\begin{bmatrix} 2 & 1 \\ 0 & 0 \end{bmatrix}$

[C] $\begin{bmatrix} 4 & 0 \\ 0 & 0 \end{bmatrix}$

[D] $\begin{bmatrix} 6 & 3 \\ 4 & 2 \end{bmatrix}$

[E] $\begin{bmatrix} 3 & 2 \\ 2 & 2 \end{bmatrix}$

The answer is: A

Multiply the rows of the first matrix by the columns of the second matrix, in each case adding the two products: $2 \times 1 + 1 \times 2 = 4$, $2 \times 1 + 1 \times 0 = 2$, $0 \times 1 + 2 \times 2 = 4$, $0 \times 1 + 2 \times 0 = 0$.

16. Find the point(s) of intersection of the graphs $f(x) = 4x^2 + 8$ and $g(x) = x^3 + 2x$.

[A] (0, 0) and (0, 8)

[B] (0, 8)

[C] $(2\sqrt{2}, 16)$

[D] (4, 72)

[E] There is no point of intersection.

The answer is: D

Algebraically, find the x-coordinate of intersection by setting the two function expressions equal to each other. $4x^2 + 8 = x^3 + 2x$

Put all terms on same side equal to zero. $0 = x^3 + 2x - 4x^2 - 8$

Factor by grouping. $0 = x(x^2 + 2) - 4(x^2 + 2)$

$0 = (x - 4)(x^2 + 2)$

Set each factor equal to zero. $x - 4 = 0$ or $x^2 + 2 = 0$

This scenario yields only one real solution for x: $x = 4$. Evaluate either function to find y: $y = 4(4)^2 + 8 = 72$. Alternatively, use graphing technology to graph the two equations and find their intersection, realizing that the intersection point exists despite the fact that it may appear outside the standard viewing window.

17. **Find the equation of the line that passes through the point (3, 7) and has a slope of $\frac{1}{3}$.**

   [A] $y = 3x + 7$

   [B] $y = \frac{1}{3}x + 7$

   [C] $y = \frac{1}{3}x + 6$

   [D] $x - 3y + 18 = 0$

   [E] Both C and D

   The answer is: E

   First find the equation of the line using point-slope form:
   $y - y_1 = m(x - x_1)$.

   $$y - 7 = \frac{1}{3}(x - 3)$$

   $$y - 7 = \frac{1}{3}x - 1$$

   $$y = \frac{1}{3}x + 6$$

   But this equation can be rewritten, first by multiplying through by 3:
   $$3y = x + 18$$
   $$0 = x - 3y + 18$$
   $$x - 3y + 18 = 0$$

   So both C and D are correct.

18. Find the zeros of the function $f(x) = x^2 - 12x - 13$.

   [A] -1

   [B] 0

   [C] 12

   [D] 13

   [E] Both A and D

   The answer is: E

   When looking at the graph of the function, the x intercepts, and therefore the zeros, are -1 and 13. Alternatively, the trinomial can be factored: $(x - 13)(x + 1)$ and show zeros of 13 and -1.

19. If $h(x) = 3x - 7$, find $-h(x)$.

   [A] $-h(x) = 3x + 7$

   [B] $-h(x) = 7 - 3x$

   [C] $-h(x) = 7x - 3$

   [D] $-h(x) = -10$

   [E] None of the above

   The answer is: B

   If $h(x) = 3x - 7$, then $-h(x) = -(3x - 7) = 7 - 3x$

20. Two friends post a video on the Internet on Monday. On Tuesday each of them forwards the video to three friends. On Wednesday, each of those friends forwards the video to three more friends. The same thing happens Thursday and Friday. How many people have seen the video by the end of Friday?

[A] 121

[B] 162.

[C] 240.

[D] 242

[E] 728

The answer is: D

The total number of viewers is
$2+(2\times 3)+(2\times 3^2)+(2\times 3^3)+(2\times 3^4)=242$.

21. Factor completely $x4 - 4x^2 - 45$

[A] $(x2 - 15)(x2 + 3)$

[B] $(x2 + 9)(x2 - 5)$

[C] $(x2 + 5)(x + 3)(x - 3)$

[D] $(x + 5)(x - 5)(x + 3)(x - 3)$

[E] None of the above

The answer is: C

First factor the trinomial in a pattern that will create a product of -45 and a sum of -4.
$$(x2 - 9)(x2 + 5)$$
Then notice that the first factor can be factored again, as a difference of two squares.
$$(x + 3)(x - 3)(x2 + 5)$$

22. The fine for an overdue library book is 50 cents on the first overdue day and increases by 5 cents on each subsequent day. How great a fine is due for a book that has been overdue for 6 days?

[A] $0.30

[B] $0.75

[C] $0.80

[D] $3.00

[E] $3.75

The answer is: B

The total fine is $.50 for the first day and 5 x $0.05 for the remaining days, adding to $0.75.

23. Which function below is not continuous over the set of real numbers?

[A] $f(x) = |x|$

[B] $g(x) = x^2$

[C] $h(x) = \dfrac{1}{x}$

[D] $q(x) = \begin{cases} 0 & \text{for } x > 10 \\ \sqrt{10-x} & \text{for } x \leq 10 \end{cases}$

[E] Both C and D are not continuous.

The answer is: C

The graph of $h(x) = \dfrac{1}{x}$ is undefined at x = 0, so the function is not continuous. While the equation listed in choice D is a piecewise function, the fact that the values at the "split" point are equal keeps the function continuous.

24. Find a polynomial function with zeros at -3, $-\sqrt{2}, 3, \sqrt{2}$.

[A] $f(x) = 3x^4 - 3x^3 + (\sqrt{2})x^2 - (\sqrt{2})x$

[B] $g(x) = x^2(x-3) + x^3(x-\sqrt{2})$

[C] $h(x) = (x^2 + 9)(x^2 + 2)$

[D] $p(x) = x^4 - 11x^2 + 18$

[E] $t(x) = x^4 - 3x^2 + \sqrt{2}$

The answer is: D

The zero of a function is the same as the root of an equation. If r is a root of a polynomial equation then $(x - r)$ is a factor. Use the 4 given zeros, or roots, to create 4 factors: $(x+3)(x-3)(x+\sqrt{2})(x-\sqrt{2})$

After multiplying the conjugate pairs: $(x^2 - 9)(x^2 - 2)$

After multiplying the binomials: $x^4 - 9x^2 - 2x^2 + 18$

which simplifies to choice D.

25. If Pv = nrt, solve for n.

[A] $n = \frac{Pv}{rt}$

[B] n = Pv − rt

[C] n = Pvrt

[D] n = P + v − r − t

[E] none of the above

The answer is: A

In order to isolate the variable n, both sides of the equation must be divided by rt. This results in choice A.

26. Which of the following expressions below is equivalent to $(x-7)^2$?

   [A] $x^2 + 49$

   [B] $x^2 - 49$

   [C] $x^2 - 14x + 49$

   [D] $49x^2$

   [E] $2x - 14$

   The answer is: C

   The squared binomial can be expanded: $(x-7)(x-7)$

   And multiplied: $x^2 - 7x - 7x + 49$

   And combined: $x^2 - 14x + 49$

27. Solve the equation $\dfrac{5a-10x}{3} = \dfrac{1}{5}$

[A] $\{47, 53\}$

[B] $\left\{\dfrac{47}{25}, \dfrac{53}{25}\right\}$

[C] $-\dfrac{47}{5}$

[D] -47

[E] 53

The answer is: B

First, isolate the absolute value bars by multiplying both sides of the equation by 3.
$$|5a-10| = \dfrac{3}{5}$$
Then set up two equations to represent the definition of absolute value.

$5a - 10 = \dfrac{3}{5}$ and $5a - 10 = -\dfrac{3}{5}$

$5a = \dfrac{53}{5}$ and $5a = \dfrac{47}{5}$

$a = \dfrac{53}{25}$ and $a = \dfrac{47}{25}$

28. Simplify $\dfrac{5a-10x}{3} = \dfrac{1}{5}$

[A] $\dfrac{x^2+3x+1}{x}$

[B] $\dfrac{x^2+11x+4}{2x+3}$

[C] $\dfrac{x^2+11x+9}{3x}$

[D] $2x + 4$

[E] 3

The answer is: A

First factor and cancel in the first portion of the expression.

$$\dfrac{x^2+11x+24}{x+8} \rightarrow \dfrac{(x+8)(x+3)}{(x+8)} \rightarrow x+3$$

Then find common denominators to add the two expressions together.

$$(x+3)+\dfrac{1}{x}$$

$$\dfrac{x+(x+3)}{x}+\dfrac{1}{x}$$

$$\dfrac{x^2+3x+1}{x}$$

29. **Select the statement below that explains the best first step to solving the equation** $\frac{3x-7}{4} = 5x+1$

[A] Add 7 to both sides.

[B] Subtract 3x from both sides.

[C] Divide both sides by 5.

[D] Multiply both sides by $\frac{1}{3}$.

[E] Multiply both sides by 4.

The answer is: E

While choices A–D do describe steps that may take place later in the equation-solving process, they are not advisable as first steps due to the interference of the denominator on the left side of the equation. Choice E as a first step "clears" that denominator so the rest of the steps can occur.

30. **Which number line shows the solution to** $7x + 5 > 9x + 17$?

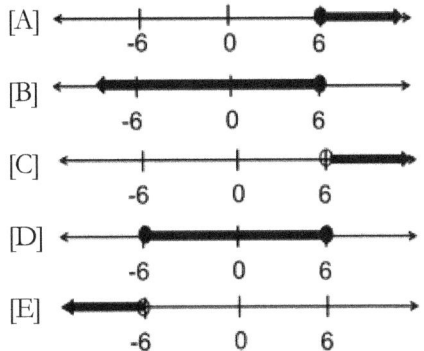

The answer is: E

First gather all the x-terms on one side of the inequality and the numbers on the other.
$7x + 5 > 9x + 17$
$-2x > 12$

Dividing both sides of an inequality by a negative number reverses the inequality sign. So division by –2 on both sides results in $x < -6$, which is graphed in choice E.

31. Subtract $3-2i$ from $10-7i$

   [A] $16i$

   [B] $7 + 5i$

   [C] $7 + 9i$

   [D] $13 + 5i$

   [E] $-7 - 5i$

   The answer is: C
   $10 + 7i - (3 - 2i)$
   $10 - 3 + 7i - (-2i)$
   $7 + 9i$

32. Given $g(x) = 2x^2 + 4$, find $g(-3)$.

   [A] -14

   [B] -8

   [C] 16

   [D] 22

   [E] 40

   The answer is: D
   $g(-3) = 2(-3)^2 + 4 = 2(9) + 4 = 22$

33. If $g(x) = x^2 + 9$ and $f(x) = x^2$, find $f(g(x))$.

   [A] $x + 3$

   [B] $2x^2 + 9$

   [C] $x^4 + 9$

   [D] $x^4 + 81$

   [E] $x^4 + 18x^2 + 81$

   The answer is: E
   Evaluate: $f(g(x)) = f(x^2 + 9) = (x^2 + 9)^2$
   Simplify by expanding: $(x^2 + 9)(x^2 + 9)$
   $$x^4 + 9x^2 + 9x^2 + 81$$
   $$x^4 + 18x^2 + 81$$

34. **Which of the following equations does not represent a function?**

   [A] $y = x^2$

   [B] $x = y$

   [C] $x = y - 5$

   [D] $x = 8$

   [E] $y = 10$

   The answer is: D

   A function can have only one output, y, for each input, x. A table of ordered pairs for choice D could be

x	8	8	8	8	8
y	-2	0	1	3	6

   This shows the (only) input, 8, has multiple outputs, y. Therefore the equation fails to be a function.

35. A canoe leaves a dock and paddles north across a river at 1.5 mph while the river current carries the canoe eastward at 2 mph. How far is the canoe from the dock after 30 minutes?

[A] 0.75 mi

[B] 1 mi

[C] 1.25 mi

[D] 1.5 mi

[E] 2 mi

The answer is: C

The motion of the canoe can be described by vectors: a northward vector of 1.5 mph created by the paddling and an eastward vector of 2 mph created by the river current.

Use the Pythagorean Theorem to find the length of the vector representing the sum of the two vectors: $1.5^2 + 2^2 = c^2$. Then c is a vector moving roughly northeast at 2.5 mph.

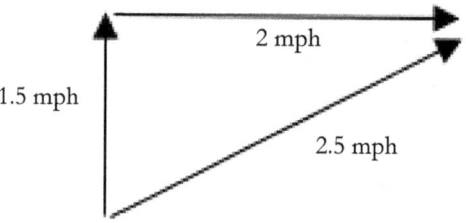

After moving away from the dock at a rate of 2.5 mph for half an hour, the canoe will be 1.25 mi distant from the dock.

36. Find the equation of a line that passes through the origin, and is parallel to the line $2x + 7y = 14$

[A] $y = -\frac{2}{7}x$

[B] $y = \frac{7}{2}x$

[C] $y = x$

[D] $x + y = 14$

[E] $x + y = 0$

The answer is: A

Find the slope of the given line by solving the equation for y.
$$2x + 7y = 14$$
$$7y = -2x + 14$$
$$y = \frac{-2}{7}x + 2$$

The slope of this line, shown now in slope intercept form, is $-\frac{2}{7}$. Any line parallel to this one must have the same slope. If the new line is to go through the origin, its y intercept is 0. These conditions yield the equation of the line $y = -\frac{2}{7}x + 0$, or choice A.

37. In the given graph, if $y_1 = x^2$, then which is the most likely equation to represent y_2?

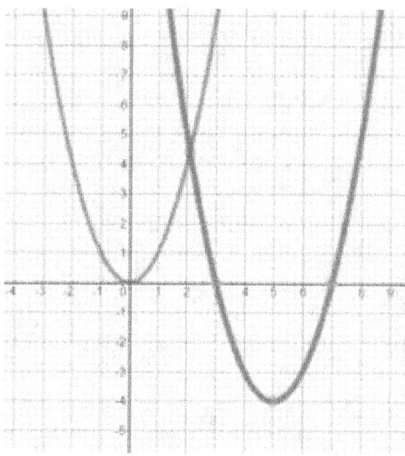

[A] $y_2 = 9x^2$

[B] $y_2 + 4 = x^2 + 5$

[C] $y_2 = (x-5)^2 - 4$

[D] $y2 = (x - 5)(x - 4)$

[E] None of the above

The answer is: C

The vertex of the graph of y_2 is (5, –4). Therefore choice C is the best equation.

38. Given the piecewise function $f(x) = \begin{cases} x+3 & \text{for } x \geq 0 \\ 5 & \text{for } x > 0 \end{cases}$ find $f(8)$.

[A] 2

[B] 5

[C] 8

[D] 11

[E] None of the above

The answer is: D

The input value, 8, is greater than zero, so the function is to be evaluated using the first portion of the rule: $f(x) = x + 3$, so $f(8) = 8 + 3 = 11$.

39. On a certain standardized test, Student A scores 50 on math, 60 on reading, and 48 on writing. Student B scores 48 on math, 45 on reading, and 55 on writing. Which matrix below best represents this data?

[A] $\begin{bmatrix} A & 50 \\ B & 48 \end{bmatrix}$

[B] $\begin{bmatrix} 50 & 60 \\ 45 & 55 \end{bmatrix}$

[C] $\begin{bmatrix} 50 & 60 & 48 \\ 48 & 45 & 55 \end{bmatrix}$

[D] $\begin{bmatrix} 50 & 60 & 48 \\ 55 & 45 & 48 \end{bmatrix}$

[E] $\begin{bmatrix} 50 & 60 & 48 \\ 75 & 60 & 50 \\ 45 & 48 & 55 \end{bmatrix}$

The answer is: C

Matrix C shows all the test data, with the rows representing each student and the columns representing math, reading, and writing.

40. Given the equation of a parabola, $y = x^2$, which equation below represents a transformation best described as a shift of 10 units up and 3 units to the left?

[A] $y = 3x^2 + 10$

[B] $y = 10(x + 3)2$

[C] $y = (x + 3)^2 + 10$

[D] $y = (x - 3)^2 - 10$

[E] $y = (x - 3)^2 + 10$

The answer is: C

The desired transformation moves the vertex of the parabola from its original (0, 0) location to the new point (-3, 10). The standard form of a parabola with vertex (h, k) is $y = (x - h)^2 + k$. Choice C, then, has the vertex (-3, 10).

41. Find the area of the region bounded by $\begin{cases} x^2 + y^2 = 25 \\ x \geq 0 \\ y \geq 0 \end{cases}$

[A] 50π

[B] 25π

[C] $\dfrac{25\pi}{4}$

[D] $\dfrac{5\pi}{4}$

[E] 25

The answer is: C

The given region is the first quadrant portion of a circle, centered on the origin, with radius 5.

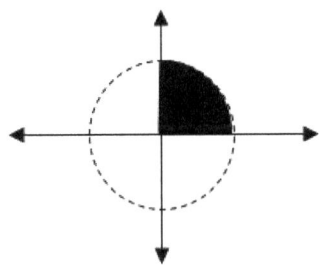

The full area of such a circle would be $A = \pi r^2 = 25\pi$. The first quadrant section is a quarter of the entire circle, resulting in answer choice C.

42. The ages of the participants in a hula-hoop contest are as follows: 10, 18, 22, 17, 77, 19, 13, 20, 10, 15. Which measure would most accurately represent the data as a whole?

[A] range

[B] interquartile range

[C] mode

[D] mean

[E] median

The answer is: E

The data set contains an outlier. The range including the outlier would represent too wide a variation of the data, most of which falls within a narrow range. The interquartile range would show that the data was mostly within a narrow range, but would indicate nothing about the general size of the data values. The mode would be a value smaller than all the other data items. The mean would be too large, skewed by the presence of the outlier. Only the median would have a value representative of the data as a whole.

43. Which equation below represents a parabola with axis of symmetry x = 3?

[A] $y^2 = (x-3)$

[B] $y = (x-3)^2$

[C] $x^2 + y^2 = 9$

[D] $y = x^2 + 3$

[E] $y = 3x^2$

The answer is: B

Choice B represents a parabola, opening up, with vertex (3, 0). Such a parabola has a line of symmetry at $x = 3$.

44. Expand $(x+2)^3$

[A] $6x$

[B] $3x + 6$

[C] $x^3 + 8$

[D] $x^3 + 3x^2 + 8$

[E] $x^3 + 6x^2 + 12x + 8$

The answer is: E

To find the polynomial answer, either multiply the binomial by itself 3 times, or apply the cube formula $(A+B)^3 = A^3 + 3A^2B + 3AB^2 + B^3$

45. The interquartile range shown in the boxplot is

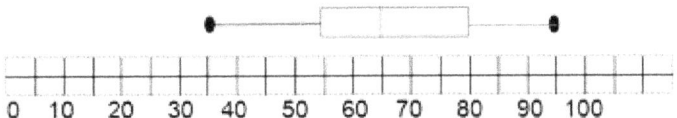

[A] 25

[B] 30

[C] 35

[D] 60

[E] 95

The answer is: A

In a box plot, the interquartile range runs from the left side of the box to the right side of the box, in this case from 55 to 80.

46. State the domain of the function $f(x) = \dfrac{3x-6}{x^2-25}$

[A] $x \neq 2$

[B] $x \neq 5, -5$

[C] $x \neq 2, -2$

[D] $x \neq 5$

[E] All real numbers

The answer is: B

The values of 5 and –5 must be omitted from the domain of all real numbers because if x took on either of those values, the denominator of the fraction would have a value of 0, and therefore the fraction would be undefined.

47. Given the function $f(x) = \begin{cases} x^2 + 5 & \text{for } x \geq 0 \\ x + 5 & \text{for } x > 0 \end{cases}$ for what values of x is the graph increasing?

[A] For $x > 0$

[B] For $x < 0$

[C] For all real values of x

[D] For $x > 5$

[E] The graph is only decreasing.

The answer is: C

When looking at the graph from left to right, it is first a line with a positive slope, making an increasing graph. The graph is continuous at $x = 0$, since the point (0, 5) is common to both equations. Then, as the graph continues to the right of the y-axis, it is the right half of a parabola, which is also an increasing graph.

48. Given the following table of values, what are possible equations for are y_1 and y_2?

x	y_1	y_2
-3	-3	-3
-2	-2	-2
-1	-1	-1
0	0	0
1	1	-1
2	2	-2
3	3	-3

[A] $, y_2 = -x$

[B] $y_1 = x, y_2 = |x|$

[C] $y_1 = x, y_2 = -|x|$

[D] $y_1 = -x, y_2 = |x|$

[E] $y_1 = |x|, y_2 = -|x|$

The answer is: C

The equation for y^1 keeps all the x-inputs the same, while the y^2 makes all the outputs negative. Choice C accomplishes these transformations.

49. In a card game, you get another turn if the card you draw is red or if it is a jack, queen, king, or ace. You are the first to draw from a full deck of 52 cards. What is the chance you will get another turn?

[A] $\dfrac{2}{13}$

[B] $\dfrac{4}{13}$

[C] $\dfrac{1}{2}$

[D] $\dfrac{17}{26}$

[E] $\dfrac{21}{26}$

The answer is: D

The cards that would give you another turn include all 26 red cards, plus the two black jacks, queens, kings, and aces, making a total of 34 out of 52 or $\dfrac{17}{26}$

50. Find the 3rd term of the polynomial resulting from the expansion (presented in descending order) of $(x + 2)^5$

[A] $7x$

[B] $8x^3$

[C] $40x^3$

[D] $32x^5$

[E] $80x^2$

The answer is: C

Use the Binomial Theorem: $(a+b)^n = \sum_{k=0}^{n} {}_nC_k a^{n-k} b^k$ where the 3rd term is represented by k=2, since k starts at zero.

$${}_5C_2 (x)^3 (2)^2 = 10(4x^3) = 40x^3$$

51. Which of the following does not represent a real number?

[A] $|14|$

[B] $\sqrt{26}$

[C] $\sqrt{-26}$

[D] $\dfrac{10}{5-5}$

[E] Both C and D

The answer is: E

While choice B is irrational, it is still real. Choice C, however, with a negative radicand, represents an imaginary, and therefore not real, number. Choice D is an undefined quotient so it is also not a representation of a real number.

52. **Simplify** $\dfrac{14x^5y^5}{7x^{-2}y}$

[A] $2x^7y^4$

[B] $7x^3y^5$

[C] $21x^7y^6$

[D] $\dfrac{1}{2x^7y^6}$

[E] $2xy$

The answer is: A

The division of 14 and 7 yields the value 2. The subtraction of exponents on the like bases (5 – (-2) = 7, 5 – 1 = 4) provides the variable portion of the answer.

53. Simplify the imaginary expression $(3i^3)^3$.

[A] $9i^6$

[B] $9i^9$

[C] $27i$

[D] $27i^9$

[E] -27

The answer is: C

First apply the outer exponent, then reduce the power of i.
$$\left(3i^3\right)^3 = 3^3 i^9 = 27i^8 i = 27\left(i^4\right)^2 i = 27i$$

54. Which expression below is equivalent to $\log_6 (25)$?

[A] $\log (19)$

[B] $\log (31)$

[C] $\log_{12} (50)$

[D] $\log_{25} (6)$

[E] $2\log_6 (5)$

The answer is: E

First rewrite the 25 as an exponential expression: $\log_6 (5^2)$. Then use the power rule for logs: $\log_b n^x = x\log_b n$. So $\log_6 (5^2) = 2\log_6 (5)$

55. Simplify $\dfrac{p!}{(p+2)!}$

[A] $\dfrac{1}{2}$

[B] $\dfrac{1}{p}$

[C] $\dfrac{1}{p+2}$

[D] $\dfrac{1}{(p+1)(p+2)}$

[E] $(p+1)(p+2)$

The answer is: D

By the definition of a factorial $(p + 2)! = (p + 2)(p + 1)(p)(p - 1)(p - 2)\ldots(1)$ or $(p + 2)(p + 1)p!$. Therefore

$$\dfrac{p!}{(p+2)(p+1)p!} = \dfrac{1}{(p+2)(p+1)}$$

56. Solve $|2x+7|<19$

[A] $-19 < x < 19$

[B] $-13 < x < 6$

[C] $x > -13$ or $x > 6$

[D] $x < -13$ or $x < 6$

[E] $x < -19$ or $x > 19$

The answer is: B

The given inequality should be set up and solved as a conjunction:

$$|2x+7|<19$$
$$-19 < 2x+7 < 19$$
$$-26 < 2x < 12$$
$$-13 < x < 6$$

57. If $3^x = 15$, then which of the following is also true?

[A] $12^x = 60$

[B] $9^x = 225$

[C] $x = 5$

[D] $x < 5$

[E] $x < 0$

The answer is: D

The relationships expressed in A and B do not follow from the given statement. To solve for x, rewrite the given equation as a logarithmic equation: $\log_3 (15) = x$. A calculator shows the value for x to be approximately 2.46, which is less than 5.

58. Solve for $x > 0$: $\log_4 (x) = 16$.

[A] 2

[B] 4

[C] 16

[D] 64

[E] 4,294,967,296

The answer is: E

The logarithmic equation can be rewritten as an exponential equation: $4^{16} = x$.

59. **Find the rule that generates the following sequence: 3, 15, 75, 375,…**

 [A] $a_n = 5n$

 [B] $a_n = 3(5n)$

 [C] $a_n = 3(5)^{n-1}$

 [D] $a_n = 3 + 5^n$

 [E] $a_n = 5^{n-1}$

 The answer is: C

 The presented sequence is geometric, with an r value, or the growth factor between each term, of 5. Therefore the rule for the sequence follows the formula $a_n = a_1(r)^{n-1}$

60. **Find the y intercept for the function $f(x) = x^2 - 4x - 12$**

 [A] -12

 [B] -6

 [C] -2

 [D] 0

 [E] 12

 The answer is: A

 The y intercept is the value, b, where the graph of the function crosses the y axis, or (0, b). Therefore find $f(0) = 0^2 - 4(0) - 12 = -12$

CLEP College Algebra
Sample Test 3

Sample Test Three

Select the choice that best answers the question

1. Which of the following expressions is not equivalent to $\dfrac{a+b}{c}$?

   [A] $\dfrac{1}{c}(a+b)$

   [B] $\dfrac{a}{c}+\dfrac{b}{c}$

   [C] $\dfrac{b+a}{c}$

   [D] $a+b \div c$

   [E] All of the above are equivalent to the given expression

2. The dosage of a certain antibiotic must be measured as 40 mg of medicine for every 25lb of patient weight. How many milligrams must be prescribed for a 140lb patient?

   [A] 5 mg

   [B] 165 mg

   [C] 180 mg

   [D] 224 mg

   [E] 250 mg

3. How many different committees can be formed by selecting 4 members from a pool of 50 candidates?

   [A] 5,527,200

   [B] 230,300

   [C] 200

   [D] 54

   [E] 4

4. Find $\sum_{n=1}^{5} n^2$

   [A] 24

   [B] 25

   [C] 55

   [D] 100

   [E] None of the above

5. **If n represents any whole number, which expression below represents the product of 2 consecutive, odd numbers?**

   [A] (2n + 1)(2n+3)

   [B] (n + 1)(n + 3)

   [C] (3n)(5n)

   [D] n(n2)

   [E] n(n+2)

6. **Simplify $2^3 i^5$**

   [A] $8i$

   [B] -8

   [C] $6i$

   [D] $-6i$

   [E] $\sqrt{-8}$

7. Find the magnitude of the vector 6i + 8j. (In other words, a vector with a horizontal component of 6 and a vertical component of 8)

   [A] $8\sqrt{3}$

   [B] 10

   [C] 14

   [D] 48

   [E] None of the above

8. Write an expression representing the following relationship: "double the square of a number."

   [A] $2n^2$

   [B] $(2n)^2$

   [C] $2(25n)$

   [D] n^4

   [E] Both A and B

9. Simplify the expression $(4x^8y^5)(4xy^3)^{-2}$

   [A] x^7y^2

   [B] $-16x^6y$

   [C] $\dfrac{x^4}{12y}$

   [D] $\dfrac{x^6}{4y}$

   [E] $\dfrac{x^6y}{4y}$

10. Solve for x: $\frac{1}{3}x + 2 = \frac{3}{5}x + \frac{1}{3}$

   [A] $\frac{2}{3}$

   [B] $\frac{4}{5}$

   [C] $\frac{25}{4}$

   [D] $\frac{35}{3}$

   [E] $\frac{48}{5}$

11. Solve over the complex numbers: $9x^2 + 49 = 0$

   [A] $\pm 2\sqrt{10}$

   [B] $\pm \frac{7i}{3}$

   [C] $\frac{7}{3}$

   [D] $\frac{49}{9}$

   [E] -40

12. Which system of equations below has an infinite number of solutions?

   [A] $\begin{cases} 5x+y=8 \\ 3x-4y=14 \end{cases}$

   [B] $\begin{cases} 2x+y=7 \\ y=4 \end{cases}$

   [C] $\begin{cases} 3x-2y=8 \\ 6x-4y=8 \end{cases}$

   [D] $\begin{cases} x+y=12 \\ 5x+5y=60 \end{cases}$

   [E] $\begin{cases} x^2+y^2=25 \\ x+y=5 \end{cases}$

13. Solve for x: $x^3 - 3x^2 - 3x + 9 = 0$

   [A] 0, 3

   [B] -2, 2, 3

   [C] $\sqrt{3}, \sqrt{3}, 3$

   [D] -3, 3, 9

   [E] 0, 3, 9

14. Solve for x. Round the answer to the nearest hundredth. $3^x = 12$

   [A] 2.26

   [B] 3.14

   [C] 4.00

   [D] 4.12

   [E] 6.00

15. Solve the compound inequality: $7 \leq 3x + 1 \leq 49$

   [A] $0 \leq x \leq 2$

   [B] $2 \leq x \leq 16$

   [C] $3.5 \leq x \leq 24.5$

   [D] $x \leq 2$ or $x \leq 16$

   [E] $x \geq 0$ or $x \leq 2$

16. Which function listed below is an example of an exponential function?

   [A] $f(x) = x^2$

   [B] $f(x) = x^3 - 3x + 1$

   [C] $f(x) = 3^x - 1$

   [D] $f(x) = \dfrac{1}{x^2}$

   [E] None of the above

17. Which function listed below is not defined for x = 0, -2?

   [A] $f(x) = x^2 + 2x$

   [B] $g(x) = \sqrt{x^2 + 2x}$

   [C] $p(x) = x(x - 2)$

   [D] $q(x) = 5$

   [E] $h(x) = \dfrac{3}{x^2 + 2x}$

18. If $h(x) = x^2$ and $g(x) = x + 3$, which statement below is false?

   [A] $g(h(x)) = x^2 + 3$

   [B] $h(g(x)) = x^2 + 9$

   [C] $(h \circ g)(x) = x^2 + 6x + 9$

   [D] $h(x) \cdot g(x) = x^3 + 3x^2$

   [E] B and D are both false

19. Which equation below represents a function with zeros -1, 2, and 7?

   [A] $f(x) = x^7 + x^2 - x$

   [B] $g(x) = x^2 + 7x - 1$

   [C] $h(x) = 2x + 7$

   [D] $m(x) = 2x^3 - x^2 + 7x$

   [E] $p(x) = x^3 - 8x^2 + 5x + 14$

20. The fees charged by a parking garage are as follows:

Hours	1	2	3	4	5
Fee	$12	$19	$26	$33	$40

   How would you summarize the fees charged?

   [A] $12 an hour

   [B] $5 plus $7 per hour

   [C] $15 an hour with a $3 discount

   [D] $4 plus $8 per hour

   [E] $3 plus $9 per hour

21. If $f(x) = \begin{cases} 2x \text{ for } x < 0 \\ \sqrt{x} \text{ for } x \geq 0 \end{cases}$, find $f(-16)$

   [A] -32

   [B] -16

   [C] 0

   [D] 4

   [E] 4i

22. If $h(x)$ and $g(x)$ are inverses of each other, then which statement below is true?

   [A] $h(x) = -g(x)$

   [B] $h(x) = \dfrac{1}{g(x)}$

   [C] $h(g(a)) = a$

   [D] $g(h(a)) = a$

   [E] Both C and D are true

23. Simplify

   [A] $\sqrt{64x^{10}y^2}$

   [B] $8x^5y\sqrt{3}$

   [C] $96x^{15}y^3$

   [D] $-512x^{15}y^3$

   [E] $\dfrac{512x^{15}}{y^3}$

24. Solve for x: $x^2 + 4x + 5 = 0$

    [A] {4, 5}

    [B] {1, 5}

    [C] {9}

    [D] {0}

    [E] No real solution

25. Solve $x - 5 = \sqrt{x+7}$

    [A] {5, 7}

    [B] {2, 9}

    [C] {9}

    [D] {0}

    [E] No real solution

26. Find the distance between the points (2, 5, -2) and (-1, 0, 4).

    [A] $\sqrt{30}$

    [B] $\sqrt{70}$

    [C] 30

    [D] 70

    [E] 100

27. Find the equation of a line that contains the point (0,6) and is perpendicular to $2x + y = 4$

   [A] $2x + y = 6$

   [B] $x + 2y = 6$

   [C] $x - 2y = -12$

   [D] $y = -6$

   [E] $x = -\dfrac{1}{4}$

28. Find the equation for the line of symmetry of the parabola $y = 2(x - 3)2 + 4$.

   [A] $y = x - 3$

   [B] $y = 4$

   [C] $y = 2$

   [D] $x = 2$

   [E] $x = 3$

29. Find the intersection point(s) of
$$\begin{cases} x^2 + y^2 = 16 \\ \dfrac{x^2}{16} + \dfrac{y^2}{9} = 1 \end{cases}$$

   [A] (16, 0)

   [B] (0, 16)

   [C] (±4, 0)

   [D] (0, ±4)

   [E] Both A and B

30. Which of the following is an equivalent representation of $\dfrac{3-4i}{1+2i}$?

[A] 3

[B] $2 - 6i$

[C] $3 - 2i$

[D] $-1 - 2i$

[E] $3 + 4i$

31. Given a circle, centered on the origin, with radius 6, which equation below represents moving that circle 3 units to the right and 5 units down?

[A] $x^2 - y^2 = 36$

[B] $3x^2 - 5y^2 = 36$

[C] $(x - 3)^2 + (y + 5)^2 = 36$

[D] $(x + 3)^2 + (y - 5)^2 = 36$

[E] $\dfrac{x^2}{9} + \dfrac{y^2}{25} = 1$

32. What is the magnitude, in Newtons, of the resulting force on an object when a horizontal force pushes the object with 23N of force while gravity exerts 10N in the vertical direction?

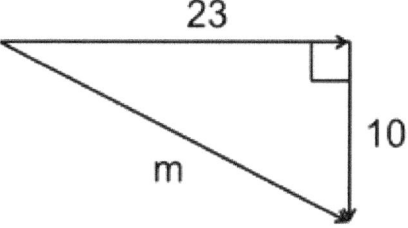

[A] 33

[B] 30.9

[C] 25.1

[D] 23

[E] 13

33. The formula for the surface area of a cube with side s is $6s^2$. The volume of a cube with side s can be calculated with the formula s^3. If the surface area of a cube is doubled, what is the resulting change in volume?

[A] The volume is doubled.

[B] The volume is increased by a factor of $2\sqrt{2}$.

[C] The volume is four times greater.

[D] The volume is eight times greater.

[E] The volume stays the same.

34. The formula for the volume of a sphere with radius r is $\frac{4}{3}\pi r^3$. If a spherical tank has a volume of 288π cm3, what is the diameter, in cm, of the sphere?

   [A] 6

   [B] 6.6

   [C] 12

   [D] 13.2

   [E] 24

35. Find the 4th term of the polynomial resulting from the expansion (presented in descending order) of $(2x + 1)^6$

   [A] $8x$

   [B] $8x^3$

   [C] $160x^3$

   [D] $200x^4$

   [E] $60x^2$

36. Find the determinant of the following matrix: $\begin{bmatrix} 3 & y \\ x & 4 \end{bmatrix}$

   [A] $7xy$

   [B] $12xy$

   [C] $x - y - 1$

   [D] $12 + xy$

   [E] $12 - xy$

37. Which choice below could correctly appear during the solving of the given equation? $15 + 3x = -8x$

   [A] $15 = -5x$

   [B] $18 = -8x$

   [C] $15 = 11x$

   [D] $15 + 11x = 0$

   [E] $5 + x = -5x$

38. Which graph represents the equation $y = 3x - x^2$?

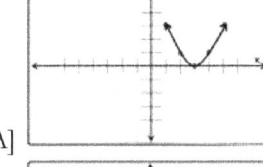
[A]

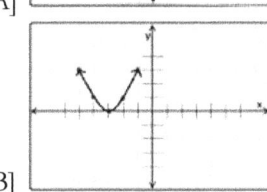
[B]

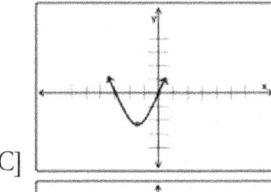
[C]

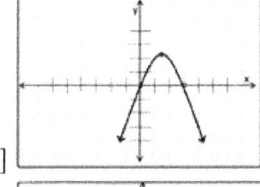
[D]

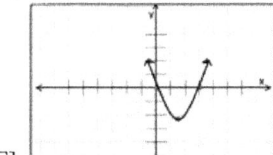
[E]

39. Solve for x $\dfrac{2}{x} + \dfrac{3}{8} = \dfrac{5}{2x}$

   [A] $\dfrac{4}{3}$

   [B] $\dfrac{8}{3}$

   [C] 4

   [D] 8

   [E] 16

40. Which of the following sets of ordered pairs does not represent a function?

   [A] {(1,4), (2,5), (3,6)}

   [B] {(1,-1), (2,-2), (3,-3)}

   [C] {(3,1), (4,1), (5,1)}

   [D] {(1,3), (1,4), (1,5)}

   [E] All of the above do represent functions

41. Which of the following is a factor of the expression $6x^2 - 5x - 14$?

   [A] $3x + 7$

   [B] $6x + 7$

   [C] $6x - 7$

   [D] $6x - 5$

   [E] $x + 2$

42. Which graph represents the solution set for $x^2 - 5x > -6$?

[A] number line with open circles at -2 and 2

[B] number line with open circles at -3 and 3

[C] number line with open circles at -2 and 2, shaded outside

[D] number line with open circles at 2 and 3

[E] number line with open circles at 2 and 3, shaded outside

43. Convert $\dfrac{7\pi}{5}$ into degrees.

[A] 4.396°

[B] 75°

[C] 175°

[D] 252°

[E] 285°

44. A group of 5 people in a room represent the following ages: 40, 32, 50, 33, and 43. If a 49 year old enters the room, which of the following will not happen?

[A] The mean age will rise.

[B] The median age will rise.

[C] The mode will not change.

[D] There will be an outlier.

[E] There will be an even number of people in the group.

45. Given a jar containing 2 red marbles, 3 white, and 8 black, what is the probability of selecting one white, replacing it, and then reaching in to select one white a second time?

[A] $\dfrac{1}{2}$

[B] $\dfrac{3}{13}$

[C] $\dfrac{6}{13}$

[D] $\dfrac{6}{169}$

[E] $\dfrac{9}{169}$

46. Given a jar containing 2 red marbles, 3 white, and 8 black, what is the probability of selecting a handful of 2 white marbles?

[A] $\dfrac{2}{3}$

[B] $\dfrac{3}{13}$

[C] $\dfrac{9}{169}$

[D] $\dfrac{3}{78}$

[E] $\dfrac{6}{169}$

47. A set of data has a mean of 78 and a standard deviation of 8. Which piece of data from the choices below is within 2 standard deviations of the mean?

[A] 16

[B] 66

[C] 93

[D] Both B and C

[E] All of the above

48. Which equation choice is a reasonable equation for the line of regression through the data points pictured below?

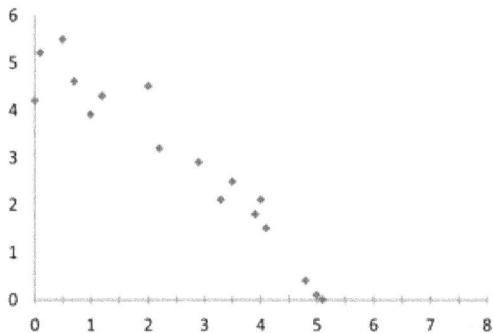

[A] $y = 5$

[B] $y = x$

[C] $y = x + 5$

[D] $y = -5x$

[E] $y = -x + 5$

49. Which of the following statements is true regarding a quadratic regression equation?

   [A] The equation can be evaluated to make predictions.

   [B] The graph of the equation does not necessarily contain all of the data points.

   [C] The graph is the line of best fit.

   [D] Both A and B are true.

   [E] All of the above are true.

50. Given the box and whisker plot below, which of the following statements regarding the plot or its corresponding data is false?

   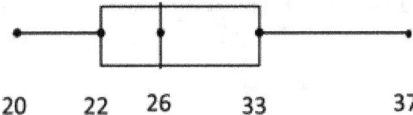

   20   22   26   33   37

   [A] The plot represents 5 pieces of data.

   [B] The median of the data is 26

   [C] The largest piece of data is 37

   [D] The interquartile range is 11

   [E] None of the statements above are false

51. Simplify $\left(9x^{16}y^{100}\right)^{1/2}$

   [A] $\dfrac{1}{9x^{16}y^{100}}$

   [B] $\dfrac{1}{18x^{32}y^{200}}$

   [C] $4.5x^8y^{50}$

   [D] $3x^8y^{50}$

   [E] $3x^4y^{10}$

52. Simplify the complex expression; $5i(3 - 2i)$

 [A] $5i$

 [B] $15i - 10$

 [C] $10 - 15i$

 [D] $10 + 15i$

 [E] $15i + 10i^2$

53. Select the expression below that is equivalent to $6!(10!)$

 [A] $4!$

 [B] $16!$

 [C] $60!$

 [D] $(6!)^2 + 4!$

 [E] $(6!)^2(7)(8)(9)(10)$

54. Select the expression below that is equivalent to $\log_n (8n^3)$

 [A] 24

 [B] $24n$

 [C] $\log (24)$

 [D] $\log_n (8) + \log_n (3n)$

 [E] $\log_n (8) + (3)$

55. Solve $|2x+7| > 19$

   [A] $-19 < x < 19$

   [B] $-13 < x < 6$

   [C] $x > -13$ or $x > 6$

   [D] $x < -13$ or $x > 6$

   [E] $x < -19$ or $x > 19$

56. Solve $5^x - 1 = 99$

   [A] 2

   [B] $\dfrac{2}{\log(5)}$

   [C] 20

   [D] $\dfrac{20}{\log(5)}$

   [E] 95

57. Solve $12 + \log_n(81) = 4\log^n(81)$ for $n > 0$.

   [A] $\dfrac{81}{4}$

   [B] $\dfrac{69}{4}$

   [C] 3

   [D] 9

   [E] None of the above

58. Which statement below is false, regarding the set of All Real Numbers, ∇?

  [A] ∇ is closed under addition.

  [B] ∇ is closed under division.

  [C] ∇ contains the set of the Irrational Numbers.

  [D] ∇ contains the set of the Imaginary Numbers.

  [E] ∇ contains an infinite number of values.

59. Find the rule that generates the sequence: 2, 9, 16, 23, ...

  [A] $a_n = n + 7$

  [B] $a_n = 2n + 7$

  [C] $a_n = 7n - 5$

  [D] $a_n = 7n + 2$

  [E] $a_n = 2(7n)$

60. Find the x intercept(s) for the function $f(x) = x^3 - 9$

  [A] -9

  [B] -3

  [C] $\sqrt[3]{9}$

  [D] 3

  [E] Both B and D

SAMPLE TEST 3 ANSWER KEY

Question Number	Correct Answer	Your Answer
1	D	
2	D	
3	B	
4	C	
5	A	
6	A	
7	B	
8	A	
9	D	
10	C	
11	B	
12	D	
13	C	
14	A	
15	B	
16	C	
17	E	
18	B	
19	E	
20	B	
21	A	
22	E	
23	E	
24	E	
25	C	
26	B	
27	C	
28	E	
29	C	
30	D	

Question Number	Correct Answer	Your Answer
31	C	
32	C	
33	B	
34	C	
35	C	
36	E	
37	D	
38	D	
39	A	
40	D	
41	B	
42	D	
43	D	
44	D	
45	E	
46	D	
47	D	
48	E	
49	D	
50	A	
51	D	
52	D	
53	E	
54	D	
55	D	
56	B	
57	C	
58	D	
59	C	
60	C	

Sample Test 3 with Rationale

1. Which of the following expressions is not equivalent to $\frac{a+b}{c}$?

   [A] $\frac{1}{c}(a+b)$

   [B] $\frac{a}{c}+\frac{b}{c}$

   [C] $\frac{b+a}{c}$

   [D] $a+b \div c$

   [E] All of the above are equivalent to the given expression

   The answer is: D

   In order for D to be an equivalent expression, $a + b$ would need to be in parenthesis

2. The dosage of a certain antibiotic must be measured as 40 mg of medicine for every 25lb of patient weight. How many milligrams must be prescribed for a 140lb patient?

   [A] 5 mg

   [B] 165 mg

   [C] 180 mg

   [D] 224 mg

   [E] 250 mg

   The answer is: D

   Set up and solve a proportion:
   $$\frac{mg \text{ of medicine}}{patient \text{ weight in lb}} \to \frac{40}{25} = \frac{x}{140}$$
   $$25x = 40(140)$$
   $$x = 224$$

3. How many different committees can be formed by selecting 4 members from a pool of 50 candidates?

[A] 5,527,200

[B] 230,300

[C] 200

[D] 54

[E] 4

The answer is: B

The combination $_{50}C_4$ represents this group of 4 people whose order does not matter. A calculator can be used to find this value; otherwise use the corresponding factorial

$$\frac{50!}{4!(50-4)!}$$

4. Find $\sum_{n=1}^{5} n^2$

[A] 24

[B] 25

[C] 55

[D] 100

[E] None of the above

The answer is: C

The given notation represents the sum of the squares of the first 5 counting numbers.

$$1^2 + 2^2 + 3^2 + 4^2 + 52^2 = 55$$

5. If n represents any whole number, which expression below represents the product of 2 consecutive, odd numbers?

[A] (2n + 1)(2n+3)

[B] (n + 1)(n + 3)

[C] (3n)(5n)

[D] n(n2)

[E] n(n+2)

The answer is: A

Since one greater than an even number is always odd, and since 2n will always be even, 2n + 1 will always be odd. The next, consecutive odd is 2 greater so 2n + 1 + 2 = 2n + 3 represents the next odd.

6. Simplify $2^3 i^5$

[A] $8i$

[B] -8

[C] $6i$

[D] $-6i$

[E] $\sqrt{-8}$

The answer is: A

The real component of the expression, 23, is 8. The imaginary portion should be broken down as follows: $i^5 = i^4 \cdot i = (1)i = i$

7. Find the magnitude of the vector $6i + 8j$. (In other words, a vector with a horizontal component of 6 and a vertical component of 8)

   [A] $8\sqrt{3}$

   [B] 10

   [C] 14

   [D] 48

   [E] None of the above

   The answer is: B

   The vector is given in terms of the standard unit vectors i and j, with a horizontal component of 6 and a vertical component of 8. Double the Pythagorean triple, 3, 4, 5, to find the hypotenuse, or vector magnitude, to be 10.

8. Write an expression representing the following relationship: "double the square of a number."

   [A] $2n^2$

   [B] $(2n)^2$

   [C] $2(25n)$

   [D] n^4

   [E] Both A and B

   The answer is: A

   The square of the number is represented by n^2, and multiplication by 2 doubles that square.

9. Simplify the expression $(4x^8y^5)(4xy^3)^{-2}$

 [A] x^7y^2

 [B] $-16x^6y$

 [C] $\dfrac{x^4}{12y}$

 [D] $\dfrac{x^6}{4y}$

 [E] $\dfrac{x^6y}{4y}$

The answer is: D

First rewrite the expression with a positive exponent:
$$\dfrac{4x^8y^5}{(4xy^3)^2} = \dfrac{4x^8y^5}{16x^2y^6} = \dfrac{x^6}{4y}$$
(Subtract exponents on like bases and reduce the fraction $\dfrac{4}{16}$)

10. Solve for x: $\frac{1}{3}x + 2 = \frac{3}{5}x + \frac{1}{3}$

[A] $\frac{2}{3}$

[B] $\frac{4}{5}$

[C] $\frac{25}{4}$

[D] $\frac{35}{3}$

[E] $\frac{48}{5}$

The answer is: C

Multiply both sides of the equation by 15, the least common denominator, to clear the fractions.

$$5x + 30 = 9x + 5$$
$$25 = 4x$$
$$x = \frac{25}{4}$$

11. **Solve over the complex numbers:** $9x^2 + 49 = 0$

    [A] $\pm 2\sqrt{10}$

    [B] $\pm \dfrac{7i}{3}$

    [C] $\dfrac{7}{3}$

    [D] $\dfrac{49}{9}$

    [E] -40

    The answer is: B

    First solve for x2, then take the plus or minus square root of both sides.
    $$9x^2 = -49$$
    $$x^2 = \dfrac{-49}{9}$$
    $$x = \pm\sqrt{\dfrac{-49}{9}} = \dfrac{7i}{3}$$

12. **Which system of equations below has an infinite number of solutions?**

    [A] $\begin{cases} 5x + y = 8 \\ 3x - 4y = 14 \end{cases}$

    [B] $\begin{cases} 2x + y = 7 \\ y = 4 \end{cases}$

    [C] $\begin{cases} 3x - 2y = 8 \\ 6x - 4y = 8 \end{cases}$

    [D] $\begin{cases} x + y = 12 \\ 5x + 5y = 60 \end{cases}$

    [E] $\begin{cases} x^2 + y^2 = 25 \\ x + y = 5 \end{cases}$

    The answer is: D

    The two equations in choice D represent the same line, as the second is the first multiplied by 5. The infinite solutions, then, are all the points along the line $x + y = 12$.

Sample Test Three

13. Solve for x: $x^3 - 3x^2 - 3x + 9 = 0$

    [A] 0, 3

    [B] -2, 2, 3

    [C] $\sqrt{3}, \sqrt{3}, 3$

    [D] -3, 3, 9

    [E] 0, 3, 9

    The answer is: C

    One method of solution is to graph the polynomial to find one real root, in this case 3, and then use synthetic division to find the remaining two solutions. This particular polynomial, however, can be factored to reveal the solutions.

    $$x^3 - 3x^2 - 3x + 9 = 0$$
    $$x^2(x-3) - 3(x-3) = 0$$
    $$(x-3)(x^2 - 3) = 0$$
    $$x - 3 = 0, \quad x^2 - 3 = 0$$
    $$x = 3, \quad x^2 = 3$$
    $$x = \pm\sqrt{3}$$

14. Solve for x. Round the answer to the nearest hundredth. $3^x = 12$

    [A] 2.26

    [B] 3.14

    [C] 4.00

    [D] 4.12

    [E] 6.00

    The answer is: A

    Rewrite the exponential equation in logarithmic form: $\log_3 12 = x$. Then estimate the log value directly on a calculator, or use the change of base rule for a calculator that only gives values of log with base 10:

    $$\log_3 12 = \frac{\log 12}{\log 3} \approx 2.26$$

15. Solve the compound inequality: $7 \leq 3x + 1 \leq 49$

   [A] $0 \leq x \leq 2$

   [B] $2 \leq x \leq 16$

   [C] $3.5 \leq x \leq 24.5$

   [D] $x \leq 2$ or $x \leq 16$

   [E] $x \geq 0$ or $x \leq 2$

   The answer is: B

   Solve the inequality by performing the same operation to all three sections.
   $$7 \leq 3x + 1 \leq 49$$
   $$6 \leq 3x \leq 48$$
   $$2 \leq x \leq 16$$

16. Which function listed below is an example of an exponential function?

   [A] $f(x) = x^2$

   [B] $f(x) = x^3 - 3x + 1$

   [C] $f(x) = 3^x - 1$

   [D] $f(x) = \dfrac{1}{x^2}$

   [E] None of the above

   The answer is: C

   An exponential function contains a variable in an exponent position.

17. **Which function listed below is not defined for $x = 0, -2$?**

   [A] $f(x) = x^2 + 2x$

   [B] $g(x) = \sqrt{x^2 + 2x}$

   [C] $p(x) = x(x - 2)$

   [D] $q(x) = 5$

   [E] $h(x) = \dfrac{3}{x^2 + 2x}$

The answer is: E

When evaluating $h(x)$ for $x = 0$ or -2, the denominator will become zero which makes the function undefined.

18. **If $h(x) = x^2$ and $g(x) = x + 3$, which statement below is false?**

   [A] $g(h(x)) = x^2 + 3$

   [B] $h(g(x)) = x^2 + 9$

   [C] $(hog)(x) = x^2 + 6x + 9$

   [D] $h(x) \cdot g(x) = x^3 + 3x^2$

   [E] B and D are both false

The answer is: B

The notation $h(g(x))$ signifies the same function composition as $(hog)(x)$. To properly calculate the function composition: $(hog)(x) = (x + 3)^2 = x^2 + 6x + 9$ when the binomial is squared correctly. Additionally, choice D correctly represents the multiplication of the two functions.

19. Which equation below represents a function with zeros -1, 2, and 7?

[A] $f(x) = x^3 + x^2 - x$

[B] $g(x) = x^2 + 7x - 1$

[C] $h(x) = 2x + 7$

[D] $m(x) = 2x^3 - x^2 + 7x$

[E] $p(x) = x^3 - 8x^2 + 5x + 14$

The answer is: E

If the zeros of the function are -1, 2, and 7, then the factors are $(x + 1)$ $(x - 2)(x - 7)$
Multiplying the three factors together yields choice E.

20. The fees charged by a parking garage are as follows:

Hours	1	2	3	4	5
Fee	$12	$19	$26	$33	$40

How would you summarize the fees charged?

[A] $12 an hour

[B] $5 plus $7 per hour

[C] $15 an hour with a $3 discount

[D] $4 plus $8 per hour

[E] $3 plus $9 per hour

The answer is: B

Choice B satisfies each entry in the table. For instance, at 4 hours the calculation is $5 + 7(4) = 33$.

21. If $f(x) = \begin{cases} 2x & \text{for } x < 0 \\ \sqrt{x} & \text{for } x \geq 0 \end{cases}$, find $f(-16)$

   [A] -32

   [B] -16

   [C] 0

   [D] 4

   [E] $4i$

   The answer is: A

   Since the input for the function is less than zero, use the top portion of the piecewise rule. $f(x) = 2x = 2(-16) = -32$

22. If $h(x)$ and $g(x)$ are inverses of each other, then which statement below is true?

   [A] $h(x) = -g(x)$

   [B] $h(x) = \dfrac{1}{g(x)}$

   [C] $h(g(a)) = a$

   [D] $g(h(a)) = a$

   [E] Both C and D are true

   The answer is: E

   If a function is applied to an input and then the inverse function is applied, the output reverts to the original input.

23. Simplify

[A] $\sqrt{64x^{10}y^2}$

[B] $8x^5y\sqrt{3}$

[C] $96x^{15}y^3$

[D] $-512x^{15}y^3$

[E] $\dfrac{512x^{15}}{y^3}$

Answer: E

Apply the exponent to each factor of the contents of the parenthesis, and multiply exponents.

$$\left(8^2\right)^{3/2}\left(x^{10}\right)^{3/2}\left(y^{-2}\right)^{3/2}$$
$$8^3 x^{15} y^{-3}$$
$$\dfrac{8^3 x^{15}}{y^3}$$

24. Solve for x: $x^2 + 4x + 5 = 0$

[A] {4, 5}

[B] {1, 5}

[C] {9}

[D] {0}

[E] No real solution

The answer is: E

As the trinomial is not factorable, the quadratic formula can be used to find a solution. The value of the discriminant, however, $b^2 - 4ac = 4^2 - 4(1)(5) = -4$ is negative, which indicates that there is no real solution to this equation.

25. Solve $x - 5 = \sqrt{x+7}$

[A] $\{5, 7\}$

[B] $\{2, 9\}$

[C] $\{9\}$

[D] $\{0\}$

[E] No real solution

The answer is: C

To solve a radical equation, raise each side to the inverse power, in this case, the power of 2.

$$(x-5)^2 = (\sqrt{x+7})^2$$
$$x^2 - 10x + 25 = x + 7$$
$$x^2 - 11x + 18 = 0$$
$$(x-9)(x-2)$$
$$x = 9, 2$$

However, when solving a radical equation, extraneous solutions can arise so the solutions must be checked in the original equation. The value -2 does not check.

$$2 - 5 \neq \sqrt{2+7}$$
$$-3 \neq \sqrt{9}$$

The 9 does check and remains the lone solution to the problem.

26. Find the distance between the points (2, 5, -2) and (-1, 0, 4).

[A] $\sqrt{30}$

[B] $\sqrt{70}$

[C] 30

[D] 70

[E] 100

The answer is: B

The distance formula for points in three dimensional space is comparable to that in two dimensions. Simply find the square root of the sum of the squares of the differences between each coordinate.

$$\sqrt{(2-(-1))^2 + (5-0)^2 + (-2-4)^2}$$
$$\sqrt{3^2 + 5^2 + (-6)^2}$$
$$\sqrt{9+25+36} = \sqrt{70}$$

27. Find the equation of a line that contains the point (0,6) and is perpendicular to $2x + y = 4$

[A] $2x + y = 6$

[B] $x + 2y = 6$

[C] $x - 2y = -12$

[D] $y = -6$

[E] $x = -\dfrac{1}{4}$

The answer is: C

Solve the given line for y to find its slope: $y = -2x + 4$. The slope is -2. Then any line perpendicular to this must have a slope of ½. With a requested y intercept of 6 and slope of ½, the new line's equation starts as $y = \dfrac{1}{2}x + 6$ which, when put in standard form, is represented by choice C.

28. **Find the equation for the line of symmetry of the parabola**
$y = 2(x-3)^2 + 4$.

[A] $y = x - 3$

[B] $y = 4$

[C] $y = 2$

[D] $x = 2$

[E] $x = 3$

The answer is: E

The given equation represents a parabola, opening up, with vertex (3, 4). A vertical line of symmetry goes through the vertex making choice E the correct answer.

29. **Find the intersection point(s) of**

$$\begin{cases} x^2 + y^2 = 16 \\ \dfrac{x^2}{16} + \dfrac{y^2}{9} = 1 \end{cases}$$

[A] (16, 0)

[B] (0, 16)

[C] (±4, 0)

[D] (0, ±4)

[E] Both A and B

The answer is: C

To find the solution algebraically, solve the first equation for y2, and substitute into the second equation.

$$y^2 = 16 - x^2$$

$$\frac{x^2}{16} + \frac{16 - x^2}{9} = 1$$

$$9x^2 + 16(16 - x^2) = 144$$

$$-7x^2 = -112$$

$$x^2 = 16$$

$$x = \pm 4$$

Confirm this solution graphically by knowing the first equation is a circle, centered on the origin, with radius 4 and the second equation represents an ellipse, also centered on the origin, with a horizontal major axis of length 8. The diameter of the circle, then, is the same as the major axis of the ellipse so the shapes intersect at each end: (4, 0) and (-4, 0).

30. Which of the following is an equivalent representation of $\dfrac{3-4i}{1+2i}$?

[A] 3

[B] 2 – 6i

[C] 3 – 2i

[D] –1 – 2i

[E] 3 + 4i

The answer is: D

Multiply both the numerator and denominator by the complex conjugate of the denominator (1 + 2i) to simplify this complex division.

$$\dfrac{3-4i}{1+2i} \cdot \dfrac{1-2i}{1-2i} = \dfrac{3-6i-4i+8i^2}{1-2i+2i-4i^2} = \dfrac{-5-10i}{1+4} = \dfrac{-5-10i}{5} = -1-2i$$

31. Given a circle, centered on the origin, with radius 6, which equation below represents moving that circle 3 units to the right and 5 units down?

[A] $x^2 - y^2 = 36$

[B] $3x^2 - 5y^2 = 36$

[C] $(x-3)^2 + (y+5)^2 = 36$

[D] $(x+3)^2 + (y-5)^2 = 36$

[E] $\dfrac{x^2}{9} + \dfrac{y^2}{25} = 1$

The answer is: C

Moving the circle as directed gives the circle a center at (3, -5). The radius remains 6. The standard form for a circle with radius r and center (h, k) is (x – h)2 + (y – k)2 = r2. Therefore, choice C is the correct answer.

32. What is the magnitude, in Newtons, of the resulting force on an object when a horizontal force pushes the object with 23N of force while gravity exerts 10N in the vertical direction?

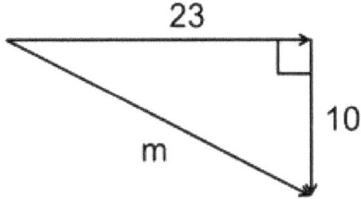

[A] 33

[B] 30.9

[C] 25.1

[D] 23

[E] 13

The answer is: C

The head-to-tail vector diagram for the given forces suggests use of the Pythagorean Theorem to solve for the magnitude of the resulting vector. $23^2 + 10^2 = m^2$, $m \approx 25.1$

33. The formula for the surface area of a cube with side s is $6s^2$. The volume of a cube with side s can be calculated with the formula s^3. If the surface area of a cube is doubled, what is the resulting change in volume?

[A] The volume is doubled.

[B] The volume is increased by a factor of $2\sqrt{2}$.

[C] The volume is four times greater.

[D] The volume is eight times greater.

[E] The volume stays the same.

The answer is: B

Start with a cube of side n, whose surface area, then, is $6n^2$ and volume is n^3. Next, double that surface area: $2(6n^2) = 12n^2$. Find the length, s, of a side of this new cube in terms of the old length, n:

$$6s^2 = 12n^2$$
$$s^2 = 2n^2$$
$$s = n\sqrt{2}$$

Then use the new length to calculate the new volume:

$$V = s^3 = (n\sqrt{2})^3 = 2n^3\sqrt{2}$$

34. The formula for the volume of a sphere with radius r is $\frac{4}{3}\pi r^3$. If a spherical tank has a volume of 288π cm3, what is the diameter, in cm, of the sphere?

[A] 6

[B] 6.6

[C] 12

[D] 13.2

[E] 24

The answer is: C

The formula for the volume of a sphere with radius r is $\frac{4}{3}\pi r^3$. Solve the equation for r.

$$V = \frac{4}{3}\pi r^3$$
$$288\pi = \frac{4}{3}\pi r^3$$
$$288\pi = \frac{4}{3}r^3$$
$$216 = r^3$$
$$r = 6$$

This means the radius is 6. Double to find the diameter of 12.

35. Find the 4th term of the polynomial resulting from the expansion (presented in descending order) of $(2x + 1)^6$

[A] $8x$

[B] $8x^3$

[C] $160x^3$

[D] $200x^4$

[E] $60x^2$

The answer is: C

Use the Binomial Theorem: $(a+b)^n = \sum_{k=0}^{n} {}_nC_k a^{n-k} b^k$ where the 4th term is represented by k=3, since k starts at zero.

$${}_6C_3(2x)^3(1)^3 = 20(8x^3)(1^3) = 160x^3$$

36. Find the determinant of the following matrix: $\begin{bmatrix} 3 & y \\ x & 4 \end{bmatrix}$

[A] $7xy$

[B] $12xy$

[C] $x - y - 1$

[D] $12 + xy$

[E] $12 - xy$

The answer is: E

To find the determinant of a 2x2 matrix, simply find the difference between the diagonal products. In this case: $3(4) - xy$

37. Which choice below could correctly appear during the solving of the given equation? $15 + 3x = -8x$

[A] $15 = -5x$

[B] $18 = -8x$

[C] $15 = 11x$

[D] $15 + 11x = 0$

[E] $5 + x = -5x$

The answer is: D

Choice D arises when $8x$ is added to both sides of the equation.

38. Which graph represents the equation $y = 3x - x^2$?

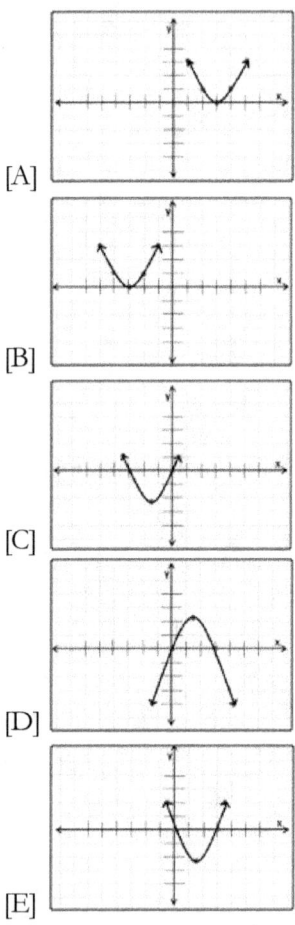

[A]

[B]

[C]

[D]

[E]

The answer is: D

Factor an x from the parabolic equation and set $y = 0$ to find the x intercepts: $0 = x(3 - x)$. The solutions for x are 0 and 3. Additionally, since the x^2 term is negative, the parabola will open downward. Therefore the answer is choice D.

39. Solve for x $\dfrac{2}{x} + \dfrac{3}{8} = \dfrac{5}{2x}$

[A] $\dfrac{4}{3}$

[B] $\dfrac{8}{3}$

[C] 4

[D] 8

[E] 16

The answer is: A

To clear the fractions, multiply both sides of the equation by the common denominator $8x$. The cleared equation is

$16 + 3x = 20$, $x = \dfrac{4}{3}$

40. Which of the following sets of ordered pairs does not represent a function?

[A] {(1,4), (2,5), (3,6)}

[B] {(1,-1), (2,-2), (3,-3)}

[C] {(3,1), (4,1), (5,1)}

[D] {(1,3), (1,4), (1,5)}

[E] All of the above do represent functions

The answer is: D

A function cannot have more than one output value (y) for the same input (x).

41. **Which of the following is a factor of the expression $6x^2 - 5x - 14$?**

[A] $3x + 7$

[B] $6x + 7$

[C] $6x - 7$

[D] $6x - 5$

[E] $x + 2$

The answer is: B

The factors $(6x + 7)(x - 2)$ correctly represent the given trinomial.

42. **Which graph represents the solution set for $x^2 - 5x > -6$?**

[A]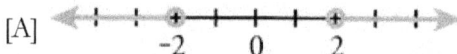
 -2   0   2

[B]
 -3   0   3

[C]
 -2   0   2

[D]
        0   2 3

[E]
        0   2 3

The answer is: D

Start by setting the quadratic inequality greater than zero:
$$x^2 - 5x + 6 > 0$$
$$(x - 2)(x - 3) > 0$$

Therefore the boundary values for the solution region are 2 and 3 but the intervals must be tested to find the regions where the trinomial is positive. Put in a number greater than 3, like 4: $4^2-5(4) + 6$, and the answer is positive. Put in a number less than 2, like 1: $1^2-5(1) + 6$, and the answer is positive. But put in a value between 2 and 3, like 2.5: $2.5^2 - 5(2.5) + 6$, and the answer is negative. This means the solutions are everywhere on the number line except for between 2 and 3, which is shown in choice D

43. Convert $\dfrac{7\pi}{5}$ into degrees.

[A] 4.396°

[B] 75°

[C] 175°

[D] 252°

[E] 285°

The answer is: D

Convert $\dfrac{7\pi}{5} rad \left(\dfrac{180 \text{ deg}}{\pi \text{ rad}} \right) = \dfrac{7}{5}(180 \text{ deg}) = 252°$

44. **A group of 5 people in a room represent the following ages: 40, 32, 50, 33, and 43. If a 49 year old enters the room, which of the following will not happen?**

[A] The mean age will rise.

[B] The median age will rise.

[C] The mode will not change.

[D] There will be an outlier.

[E] There will be an even number of people in the group.

The answer is: D

Since the number 49 is greater than both the mean and median of the original set of data, both A and B are true. And since no values are repeated either before or after the 49 year old enters, choice C is true. Assuming there is data to represent each group member, choice E is correct. Choice D is not true, because the value 49 is reasonably close to the existing data. No extremes or outliers exist.

45. Given a jar containing 2 red marbles, 3 white, and 8 black, what is the probability of selecting one white, replacing it, and then reaching in to select one white a second time?

[A] $\dfrac{1}{2}$

[B] $\dfrac{3}{13}$

[C] $\dfrac{6}{13}$

[D] $\dfrac{6}{169}$

[E] $\dfrac{9}{169}$

The answer is: E
The probability of selecting one white marble is 3 out of 13, or $\dfrac{3}{13}$. To repeat this as an independent event is to multiply the probabilities. $\dfrac{3}{13} \cdot \dfrac{3}{13} = \dfrac{9}{169}$

46. Given a jar containing 2 red marbles, 3 white, and 8 black, what is the probability of selecting a handful of 2 white marbles?

[A] $\dfrac{2}{3}$

[B] $\dfrac{3}{13}$

[C] $\dfrac{9}{169}$

[D] $\dfrac{3}{78}$

[E] $\dfrac{6}{169}$

The answer is: D
To grab a handful is to calculate combinations. How many ways can 2 whites out of the 3 be chosen? How many ways can a handful of 2 marbles be chosen out of the total 13? $\dfrac{_3C_2}{_{13}C_2} = \dfrac{3}{78}$

262   CLEP Algebra

47. A set of data has a mean of 78 and a standard deviation of 8. Which piece of data from the choices below is within 2 standard deviations of the mean?

[A] 16

[B] 66

[C] 93

[D] Both B and C

[E] All of the above

The answer is: D

A standard deviation, σ, away from the mean can go in the positive as well as the negative direction. As illustrated in the number line, both 66 and 93 are within 2 standard deviation lengths of the mean.

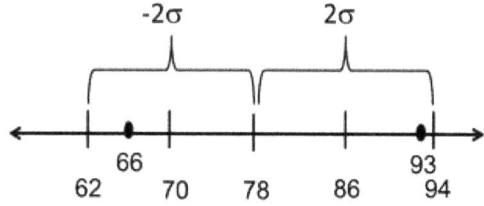

48. Which equation choice is a reasonable equation for the line of regression through the data points pictured below?

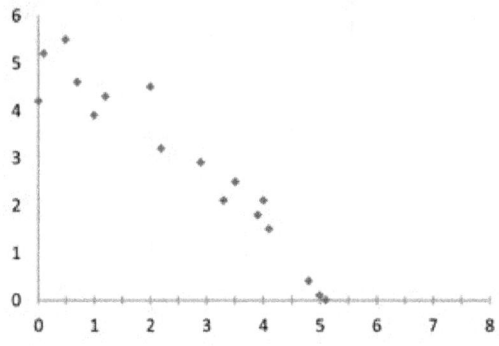

[A] $y = 5$

[B] $y = x$

[C] $y = x + 5$

[D] $y = -5x$

[E] $y = -x + 5$

The answer is: E

The best fit line drawn through the data points will have a negative slope and have intercepts close to (0, 5) and (5, 0). This makes choice E the best approximation for the regression line out of the choices given, as it contains the point (0, 5) and has a slope of -1.

49. **Which of the following statements is true regarding a quadratic regression equation?**

   [A] The equation can be evaluated to make predictions.

   [B] The graph of the equation does not necessarily contain all of the data points.

   [C] The graph is the line of best fit.

   [D] Both A and B are true.

   [E] All of the above are true.

   The answer is: D

   Choice C is not true for a quadratic regression equation as the line of best fit refers to a *linear* regression equation. A quadratic regression equation can be described as the best fit *curve* through the points. It is drawn through most of the points but does not always contain every data point. Once the equation is determined, it can be used to make predictions as to where new data points might fall.

50. Given the box and whisker plot below, which of the following statements regarding the plot or its corresponding data is false?

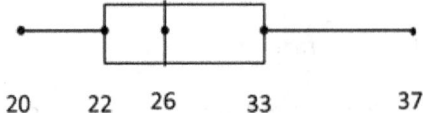

20   22   26   33   37

[A] The plot represents 5 pieces of data.

[B] The median of the data is 26

[C] The largest piece of data is 37

[D] The interquartile range is 11

[E] None of the statements above are false

The answer is: A

The box and whisker plot is built as a number line highlighting the minimum and maximum of the data (here seen as 20 and 37, respectively), as well as the three quartile values (in this case 22, 26, and 33). An indeterminate number of data pieces exist to create this spread that are not shown on the box and whisker plot. We only know that half of the data, for instance, lies between quartile 1 and 3, but we do not know how many pieces of data there are.

51. Simplify $\left(9x^{16}y^{100}\right)^{1/2}$

[A] $\dfrac{1}{9x^{16}y^{100}}$

[B] $\dfrac{1}{18x^{32}y^{200}}$

[C] $4.5x^8y^{50}$

[D] $3x^8y^{50}$

[E] $3x^4y^{10}$

The answer is: D

Since the parenthesis contain a monomial expression, the power can be distributed to each factor.

$$9^{1/2}\left(x^{16}\right)^{1/2}\left(y^{100}\right)^{1/2} = \sqrt{9}\left(x^8\right)\left(y^{50}\right) = 3x^8y^{50}$$

266   CLEP Algebra

52. Simplify the complex expression; $5i(3 - 2i)$

    [A] $5i$

    [B] $15i - 10$

    [C] $10 - 15i$

    [D] $10 + 15i$

    [E] $15i + 10i^2$

    The answer is: D

    Begin by distributing the $5i$ and multiplying: $5i(3) - 5i(2i) = 15i - 10i^2$
    Then reduce the power of i since $i^2 = -1$: $15i - 10(-1) = 10 + 15i$

53. Select the expression below that is equivalent to 6!(10!)

    [A] 4!

    [B] 16!

    [C] 60!

    [D] $(6!)^2 + 4!$

    [E] $(6!)^2(7)(8)(9)(10)$

    The answer is: E

    By the definition of factorials, $10! = 10 \cdot 9 \cdot 8 \cdot 7 \cdot 6 \cdot 5 \cdot \ldots \cdot 1 = (10 \cdot 9 \cdot 8 \cdot 7)6!$
    Therefore, $6!(10!) = 6!(10 \cdot 9 \cdot 8 \cdot 7)6!$ which then leads to choice E

54. Select the expression below that is equivalent to $\log_n (8n^3)$

   [A] 24

   [B] 24n

   [C] log (24)

   [D] $\log_n (8) + \log_n (3n)$

   [E] $\log_n (8) + (3)$

   The answer is: D
   First separate the factors by using the log rule $\log_b (ac) = \log_b (a) + \log_b (c)$
   $\log_n (8n^3) = \log_n (8) + \log_n (n^3)$
   Then use the exponent log rule: $\log_b (a^d) = d \cdot \log_b (a)$
   $\log_n (8) + \log_n (n^3) = \log_n (8) + 3\log_n (n) = \log_n (8) + 3$ since $\log_n (n) = 1$

55. Solve $|2x+7| > 19$

   [A] $-19 < x < 19$

   [B] $-13 < x < 6$

   [C] $x > -13$ or $x > 6$

   [D] $x < -13$ or $x > 6$

   [E] $x < -19$ or $x > 19$

   The answer is: D
   Set up and solve a disjunction: $2x + 7 < -19$ or $2x + 7 > 19$
   $\qquad\qquad\qquad\qquad\qquad 2x < -26$ or $2x > 12$
   $\qquad\qquad\qquad\qquad\qquad x < -13$ or $x > 6$

56. Solve $5^x - 1 = 99$

 [A] 2

 [B] $\dfrac{2}{\log(5)}$

 [C] 20

 [D] $\dfrac{20}{\log(5)}$

 [E] 95

The answer is: B

Isolate the term with the exponent, then rewrite as a logarithmic equation.
$$5^x = 100, \text{ so } \log_5(100) = x$$
Next, rewrite the logarithmic value in base 10:
$$\log_5(100) = \frac{\log(100)}{\log(5)} = \frac{2}{\log(5)}$$

57. Solve $12 + \log_n(81) = 4\log^n(81)$ for n > 0.

 [A] $\dfrac{81}{4}$

 [B] $\dfrac{69}{4}$

 [C] 3

 [D] 9

 [E] None of the above

The answer is: C

Since the logarithmic expressions are of the same base and input, they can be combined as like terms, albeit from opposite sides of the equation. Subtract $\log_n(81)$ from both sides to get:
$12 = 3\log_n(81)$
$4 = \log_n(81)$ after division of 3 on both sides
Rewrite as an exponential expression: $n^4 = 81$ and conclude that n must equal 3.

58. Which statement below is false, regarding the set of All Real Numbers, ∇?

[A] ∇ is closed under addition.

[B] ∇ is closed under division.

[C] ∇ contains the set of the Irrational Numbers.

[D] ∇ contains the set of the Imaginary Numbers.

[E] ∇ contains an infinite number of values.

The answer is: D

The Imaginary Numbers represent values derived from the square root of negative one and are distinctly separate from the set of Real Numbers.

59. Find the rule that generates the sequence: 2, 9, 16, 23, ...

[A] $a_n = n + 7$

[B] $a_n = 2n + 7$

[C] $a_n = 7n - 5$

[D] $a_n = 7n + 2$

[E] $a_n = 2(7n)$

The answer is: C

The sequence is arithmetic with a common difference, d, of 7. The rule, therefore, follows the pattern $a_n = a_1 + d(n - 1)$ which becomes $2 + 7(n - 1)$ for the given example.

Simplification yields choice C: $2 + 7n - 7 = 7n - 5$.

60. **Find the x intercept(s) for the function $f(x) = x^3 - 9$**

   [A] -9

   [B] -3

   [C] $\sqrt[3]{9}$

   [D] 3

   [E] Both B and D

   The answer is: C

   An x intercept is the point where the graph of the function crosses the x axis. Therefore y has a value of 0 at this point. So $f(x) = y = 0 = x^3 - 9$. Solving for x yields choice C.

XAMonline
The CLEP Specialist
Individual Sample Tests in ebook format with full explanations

eBooks

All 33 CLEP sample tests are available as ebook downloads from retail websites such as **Amazon.com** and **Barnesandnoble.com**

American Government	9781607875130
American Literature	9781607875079
Analyzing and Interpreting Literature	9781607875086
Biology	9781607875222
Calculus	9781607875376
Chemistry	9781607875239
College Algebra	9781607875215
College Composition	9781607875109
College Composition Modular	9781607875437
College Mathematics	9781607875246
English Literature	9781607875093
Financial Accounting	9781607875383
French	9781607875123
German	9781607875369
History of the United States I	9781607875178
History of the United States II	9781607875185
Human Growth and Development	9781607875444
Humanities	9781607875147
Information Systems	9781607875390
Introduction to Educational Psychology	9781607875451
Introductory Business Law	9781607875420
Introductory Psychology	9781607875154
Introductory Sociology	9781607875352
Natural Sciences	9781607875253
Precalculus	9781607875345
Principles of Macroeconomics	9781607875406
Principles of Microeconomics	9781607875468
Principles of Marketing	9781607875475
Principles of Management	9781607875468
Social Sciences and History	9781607875161
Spanish	9781607875116
Western Civilization I	9781607875192
Western Civilization II	9781607875208

TO ORDER  or  or

XAMonline
CLEP
Full Study Guides

CLEP College Algebra
ISBN: 9781607875598
Price: $34.95

CLEP Biology
ISBN: 9781607875314
Price: $34.95

CLEP Analyzing and
Interpreting Literature
ISBN: 9781607875260
Price: $34.95

CLEP College Composition
and Modular
ISBN: 9781607875277
Price: $19.99

CLEP College Mathematics
ISBN: 9781607875321
Price: $34.95

CLEP Psychology
ISBN: 9781607875291
Price: $34.95

CLEP Spanish
ISBN: 9781607875284
Price: $34.95

TO ORDER  or  or

XAMonline
CLEP Subject Series
Collection by Topic
Sample Test Approach

CLEP Literature
ISBN: 9781607875833
Price: $34.95

CLEP Foreign Language
ISBN: 9781607875772
Price: $34.95

CLEP History
ISBN: 9781607875789
Price: $34.95

CLEP Sociology
ISBN: 9781607875796
Price: $34.95

CLEP Science
ISBN: 9781607875802
Price: $34.95

CLEP Mathematics
ISBN: 9781607875819
Price: $34.95

CLEP Business
ISBN: 9781607875826
Price: $34.95

  or amazon or

XAMonline

CLEP Favorites

Collection by Topic
Sample Test Approach

CLEP Five Favorites
ISBN: 9781607875765
Price: $24.95

CLEP Military Favorites
ISBN: 9781607875512
Price: $24.95

TO ORDER  or  or

www.ingramcontent.com/pod-product-compliance
Lightning Source LLC
Chambersburg PA
CBHW060946230426
43665CB00015B/2083